Lecture Notes in Computer Science

T0238376

Commenced Publication in 1973
Founding and Former Series Editors:
Gerhard Goos, Juris Hartmanis, and Jan van Leeuwen

Editorial Board

David Hutchison
  Lancaster University, UK
Takeo Kanade
  Carnegie Mellon University, Pittsburgh, PA, USA
Josef Kittler
  University of Surrey, Guildford, UK
Jon M. Kleinberg
  Cornell University, Ithaca, NY, USA
Alfred Kobsa
  University of California, Irvine, CA, USA
Friedemann Mattern
  ETH Zurich, Switzerland
John C. Mitchell
  Stanford University, CA, USA
Moni Naor
  Weizmann Institute of Science, Rehovot, Israel
Oscar Nierstrasz
  University of Bern, Switzerland
C. Pandu Rangan
  Indian Institute of Technology, Madras, India
Bernhard Steffen
  University of Dortmund, Germany
Madhu Sudan
  Massachusetts Institute of Technology, MA, USA
Demetri Terzopoulos
  University of California, Los Angeles, CA, USA
Doug Tygar
  University of California, Berkeley, CA, USA
Gerhard Weikum
  Max-Planck Institute of Computer Science, Saarbruecken, Germany

Peter P. Chen   Leah Y. Wong (Eds.)

Active Conceptual Modeling of Learning

Next Generation Learning-Base
System Development

 Springer

Volume Editors

Peter P. Chen
Louisiana State University
Computer Science Department
Baton Rouge, LA 70803, USA
E-mail: chen@csc.lsu.edu

Leah Y. Wong
Space and Naval Warfare Systems Center
San Diego Command and Control Department
San Diego, CA 92152, U.S.A.
E-mail: leah.wong@navy.mil

Library of Congress Control Number: 2007941872

CR Subject Classification (1998): H.2, H.4, F.4, I.2.4, H.1, J.1, D.2, C.2

LNCS Sublibrary: SL 3 – Information Systems and Application, incl. Internet/Web
and HCI

ISSN   0302-9743
ISBN-10   3-540-77502-1 Springer Berlin Heidelberg New York
ISBN-13   978-3-540-77502-7 Springer Berlin Heidelberg New York

This work is subject to copyright. All rights are reserved, whether the whole or part of the material is
concerned, specifically the rights of translation, reprinting, re-use of illustrations, recitation, broadcasting,
reproduction on microfilms or in any other way, and storage in data banks. Duplication of this publication
or parts thereof is permitted only under the provisions of the German Copyright Law of September 9,
1965, in its current version, and permission for use must always be obtained from Springer. Violations are
liable to prosecution under the German Copyright Law.

Springer is a part of Springer Science+Business Media

springer.com

© Springer-Verlag Berlin Heidelberg 2007

Typesetting: Camera-ready by author, data conversion by Scientific Publishing Services, Chennai, India
Printed on acid-free paper
SPIN: 12210466   06/3180   5 4 3 2 1 0

Preface

This volume contains a collection of the papers presented during the 1st International ACM-L Workshop, which was held on November 8, 2006 during the 25th International Conference on Conceptual Modeling, ER 2006, held November 6–9, 2006, in Tucson, Arizona, plus several invited papers. From 26 papers submitted to the workshop, a total of 11 papers were selected for inclusion in this volume. These papers plus the invited papers represent the current thinking in conceptual modeling research.

To achieve the ACM-L goals, we need the participation of many research groups with different skill sets. The active model can only be realized through technology integration (e.g., AI, software engineering, information technology, cognitive science, art and sciences, philosophy, etc.) and by combining related modeling techniques. The current state of the art in conceptual modeling can be used as the starting point. Modeling techniques will need to be combined to model different characteristics of the world (e.g., temporal, multimedia, spatial, cognitive, philosophical, historical, etc.) in a holistic and visualized manner.

One of the suggested action items of the workshop participants was to have more workshops and working sessions on this topic. The first International ACM-L Workshop was organized to follow on from a smaller workshop on the same subject held several months earlier. This volume includes the paper presentations, panel discussions, research findings, and future directions. The organization of the workshop involved many colleagues and valuable contributions from a number of persons. We would like to thank the ER 2006 Conference Co-chair, Sudha Ram (University of Arizona, USA) and Mohan R. Tanniru (University of Arizona, USA), the Workshop and Tutorial Chair, John Roddick (Flinders University, Australia), Publicity Chair and Webmaster, Huimin Zhao (University of Wisconsin, Milwaukee, USA), and Local Arrangements and Registration, Anji Siegel (University of Arizona, USA) for their support. We would like to especially express our thanks to members of the Program Committee, external referees, and workshop session chairs, who contributed to the success of the workshop. We also would like to thank Arthur Shr of LSU, who directed a group of student helpers to assist in editing this volume.

November 2006

P.P. Chen
L. Wong

Workshop Organization

Workshop Chair

Leah Wong (Space and Naval Warfare Systems Center, San Diego, USA)

Workshop Honorary Chair

Peter Chen (Louisiana State University, USA)

Program Committee

Jackie Akoka	INT, France
Dave Embley	Brigham Young University, USA
Ray Liuzzi	Raymond Technologies, USA
Sudha Ram	University of Arizona, USA
Peter Scheuermann	Northwestern University, USA
Ted Senator	DARPA, USA
Timothy K. Shih	Tamkang University, Taiwan
Bernhard Thalheim	Christian-Albrechts-Universität zu Kiel, Germany
T.C. Ting	University of Connecticut, USA

External Referees

Gove Allen	Tulane University, USA
Ye-Sho Chen	Louisiana State University, USA
Jianhua Chen	Louisiana State University, USA
Hannu Kangassalo	University of Tampere, Finland
Stephen W. Liddle	Brigham Young University, USA
Tok Wang Ling	National University of Singapore, Singapore
Ashok Malhotra	Oracle, USA
Salvatore T. March	Vanderbilt University, USA
Il-Yeol Song	Drexel University, USA

Session Chairs

Susan D. Urban	Arizona State University, USA
Richard T. Snodgrass	University of Arizona, USA

Table of Contents

Invited Paper

Overview of Papers in 2006 Active Conceptual Modeling of Learning (ACM-L) Workshop

Peter P. Chen[1,*] and Leah Y. Wong[2,**]

[1] Computer Science Department, Louisiana State University
Baton Rough, LA 70803, U.S.A.
[2] Space and Naval Warfare Systems Center San Diego
Code 246214, 53560 Hull Street
San Diego, CA 92152, U.S.A.
pchen@lsu.edu, leah.wong@navy.mil

Abstract. This is a summary of the papers presented at the Active Conceptual Modeling of Learning (ACM-L) Workshop, November 8, 2006, Tucson, Arizona, USA, and several invited papers.

1 Introduction

As a result of the call for papers, the Program Committee received 26 submissions from 10 countries, and after rigorous refereeing 11 papers were eventually chosen for presentation at the workshop. The workshop program was enriched by a keynote address given by Professor Peter Chen, the Honorary Chair of the workshop. The first part of this paper is a summary of the papers in the order of the presentations at the workshop. The second part describes the invited papers.

2 Summary of Regular Papers

Understanding the Semantics of Data Provenance to Support Active Conceptual Modeling
Sudha Ram and Jun Liu, University of Arizona, Tucson, AZ, USA

The paper provides a comprehensive approach for recording how data changes overtime based on the W7 model conceptualizing dimensions of data provenance. The authors introduce how data provenance could generalize the data-centric and process-centric approaches by using ontology to represent semantics of provenance.

---

* The research of this author was partially supported by National Science Foundation grant: NSF-IIS-0326387 and Air Force Office of Scientific Research (AFOSR) grant: FA9550-05-1-0454.
** Contributions to this article by co-author Leah Wong constitute an original work prepared by an officer or employee of the United States Government as part of that person's official duties. No copyright applies to these contributions.

P.P. Chen and L.Y. Wong (Eds.): ACM-L 2006, LNCS 4512, pp. 1–6, 2007.
© Springer-Verlag Berlin Heidelberg 2007

They suggest that tracking data provenance could enable users to share, discover, and reuse data, thus streamlining collaborative activities and facilitating learning. The authors propose to use the provenance-based conceptual modeling approach to capture "what" events/changes have happened to the data, identify "where", "when", "how", "who", and "why" behind the "what"; and represent events as entities with attributes and relationships without adding additional constructs. The paper also uses a homeland security example to illustrate how current conceptual models can be extended to embed provenance in order to provide integrated information in a holistic manner.

Adaptive and Context-Aware Reconciliation of Reactive and Pro-Active Behavior in Evolving Systems
Peter Scheuermann and Goce Trajcevski, Northwestern University, USA

The authors introduce the $(ECA)^2$ model, which extended active databases and focused on streaming data, for modeling reactive behavior with pro-active impact of evolving systems including hypothetical reasoning, dynamic trigger reordering, and context-aware adaptation. Proactive impact of the reactive behavior includes:

1. Modification of the events/conditions/actions as the system under consideration evolves
2. Dynamic ordering of the execution of the "awaked" triggers
3. Push-based notification of the satisfaction of the continuous queries (condition part) of the triggers.

A Common Core for Active Conceptual Modeling for Learning from Surprises
Stephen W. Liddle and David W. Embley, Brigham Young University, USA

This paper introduces a formal foundation for active conceptual modeling in terms of meta-modeling, formal representation, and mathematical model theory based on the Object-oriented Systems Modeling (OSM) approach with examples of executable models, and a set of suggested common ACM-L core concepts for representing the structural and dynamic aspects of a system. The authors call the ACM-L community to agree on a common core model that supports all aspects—static and dynamic— needed for active conceptual modeling in support of learning from surprises. It is the authors' position that we have the building blocks for realizing ACM-L, but they need to be put together in a suitable way.

Actively Evolving Conceptual Models for Mini-world and Run-time Environment Changes
P. Radha Krishna and Kamalakar Karlapalem, Institute for Development and Research in Banking Technology, Hyderabad, India

This paper introduces an ER* methodology for evolving applications that implicitly handle changes and synchronization at various levels from the mini-world to run-time environments through layers of models and systems. A suite of ER Models, which act as a template for modeling the change requirements during application evolution, is used to describe various requirements. An ER* model will be instantiated by invoking the most appropriate template that best meets the requirements of the changed environment. The methodology is iterative in nature and is driven by the structured, functional, and behavior validation steps, along with the Event-Condition-Action (ECA) rules for monitoring and execution of an application. This methodology facilitates capturing active behavior from run-time transactions and provides a means of using the knowledge to guide subsequent application design and its evolution. An e-contract was provided as an example.

Achievements and Problems of Conceptual Modeling
Bernhard Thalheim, University Kiel, Olshausenstrasse Kiel, Germany

This paper introduces state-of-the-art and open problems in structuring, functionality (behavior), advanced views and media types, distribution, and interactivity in conceptual modeling. Twenty open problems with respect to the achievements of conceptual modeling were introduced. These achievements, which demonstrated the maturity towards a science of modeling, have laid a foundation for database technology in support of a large class of applications. The open problems include issues in database design, information architecture, information quality, specification language development, conceptual modeling theory, constraints and consistency management, theory of extended operations for object-oriented models and database behaviors, mapping to triggers, and adaptable delivery of content to the user.

Metaphor Modeling on the Semantic Web
Bogdan D. Czejdo, CASS, San Antonio, TX, USA, Jonathan Biguenet, Tulane University, USA, and John Biguenet, Loyola University, USA

This paper introduces the use of the extended Unified Modeling Language (UML) for metaphor modeling. It further discussed how to create UML diagrams to capture knowledge about metaphors for defining complex meaning. The phases of metaphor modeling include stating the metaphor, identifying similarity of attribute types, the role of matching attributes for a metaphor, and explaining the non-matching attribute role for the metaphor. The authors suggested application of metaphor-based systems in education and surprise detection. However, the changes of metaphor and the underlying ontology remained challenging issues that need to be addressed.

Schema Changes and Historical Information in Conceptual Models in Support of Adaptive Systems
Luqi, Naval Postgraduate School, USA and Douglas S. Lange, Space and Naval Warfare Systems Center San Diego, USA

The authors point out software engineering, which is an example of learning activity, could be used as an ACM-L use case capturing the essence of version control, software merge, reuse, and generation. While mapping changes of the real world domain to software world requires theory and methodologies, this paper describes several aspects of active conceptual modeling that have been studied in software engineering and suggests further investigation when applying ER models to active conceptual learning. Specifically, uncertainty and time models need to be incorporated.

Using Active Modeling in Counterterrorism
Yi-Jen Su, Hewijin C. Jiau, and Shang-Rong Tsai, National Cheng Kung University, Taiwan

This paper proposes using active modeling in analyzing unconventional attacks in the design of counterterrorism system. The authors combined terrorist network analysis and active modeling for an Intelligent Terrorism Detection System including intelligence evaluation and information binding. It uses the Threat Factor and Threat Pattern Finding (TFIDF) to classify the category of potential and Graph-Based Mining to recognize the frequent threat patterns from the past incidents. Inexact graph matching is then applied to detect the potential threat from suspicious terrorist networks. Terrorism ontology is used to capture the semantic content of the terrorist domain. Based on time sequence, snapshots are used to describe the activity sequence of a terrorist attack.

To Support Emergency Management by Using Active Modeling: A Case of Hurricane Katrina
Xin Xu, Louisiana State University, USA

The author discusses the issues of how to deal with surprises and determine changes, which lead to influence relationships among entities using a case study of Hurricanes Katrina. Current work in disaster process modeling needs to be enhanced to include a conceptual model for emergency management by integrating data from historical information (experience, crisis, and surprise), and representing relationship from all aspects of a domain in a dynamic way. The paper proposes a process framework that would help predict surprises within upcoming natural disaster. This framework assumes all types of natural disasters have occurred, and corresponding information, experience, and lessons have been classified and stored; and that we only focus on the major upcoming disaster. Specific issues that need to be further explored include: (1) How to capture the right information from changes during emergency situation, (2) How to adapt to new situation by learning from past experience, (3) How to predict potential changes by integrating current and past knowledge.

Using Ontological Modeling in a Context-Aware Summarization System to Adapt Text for Mobile Devices
Luís Fernando Fortes Garcia, Lutheran University of Brazil, Brazil, Faculdade Dom Bosco de Porto Alegre, Brazil, José Valdeni de Lima, Federal University of Rio Grande do Sul, Brazil, Stanley Loh, Lutheran University of Brazil, Brazil, Catholic University of Pelotas, Brazil, and José Palazzo Moreira de Oliveira, Federal University of Rio Grande do Sul, Brazil

The authors introduce context awareness for dealing with linking changes in the environments with computer systems, which are otherwise static. The paper illustrates the use of ontological modeling for context-aware summarization to present users' interests, and spatial and temporal localizations in order to adapt text for mobile devices. The paper proposed the use of contextual information to determine "what", "where", and "when" should be delivered to the user in different environments according to the user's profile. Domain ontologies are structured as concept hierarchies with each concept being described by a set of keywords with weight assignment to determine how much the word identifies the concept, which in turn derives the summarization. An open architecture consisting of the user's profile management module, context management module, and context-aware summarization modules and automated text summarization process based on context and personal information is presented. Results of experimentations are also presented.

Accommodating Streams to Support Active Conceptual Modeling of Learning From Surprises
Subhasish Mazumdar, New Mexico Institute of Mining and Technology, USA

This paper proposes an enhancement of ER modeling with active constructs in order to permit data streams, which can only be queried by context, to have context-based relationship with standard data.

3 Summary of Invited Papers

Approaches to the Active Conceptual Modelling of Learning
Hannu Kangassalo, University of Tampere, Finland

This paper studies that collection on several levels of abstraction of human cognition and knowledge. These processes can be performed through various approaches, on several levels, and by using several perspectives. The paper concentrates on active conceptual modelling, which is a process of recognition, finding or creating relevant concepts and conceptual models which describe the UoD, representing the conceptual content of information to be contained in the IS. This characterisation contains the construction of new concepts, too. The author studies methods for collecting information from various sources in the UoD and accumulating it as possibly actual instances of various types of pre-defined concepts. Some of these instances may be cases of sudden events or processes. They should be recognised as concepts and included in to the conceptual schema. To some extent, some concepts may be constructed which fit to this collected information. During the adaptation process it is recommended that we are applying active conceptual modelling for learning, which organises our conceptual schema in a new way. Learning is a process in which a learner re-organises, removes or refills his knowledge structures by applying his newly organised conceptual schema.

Spatio-Temporal and Multi-Representation Modeling: A Contribution to Active Conceptual Modeling
Stefano Spaccapietra, Ecole Polytechnique Fédérale de Lausanne, Switzerland, Christine Parent, University of Lausanne, Switzerland, and Esteban Zimányi, Université Libre de Bruxelles, Belgium

Worldwide globalization increases the complexity of problem solving and decision-making, whatever the endeavor is. This calls for a more accurate and complete understanding of underlying data, processes and events. Data representations have to be as accurate as possible, spanning from the current status of affairs to its past and future statuses, so that it becomes feasible, in particular, to elaborate strategies for the future based on an analysis of past events. Active conceptual modeling is a new framework intended to describe all aspects of a domain. It expands the traditional modeling scope to include, among others, the ability to memorize and use knowledge about the spatial and temporal context of the phenomena of interest, as well as the ability to analyze the same elements under different perspectives. This paper shows how these advanced modeling features are provided by the MADS conceptual model.

Postponing Schema Definition: Low Instance-to-Entity Ratio (LItER) Modelling
John F. Roddick, Aaron Ceglar, Denise de Vries and Somluck La-Ongsri, Flinders University, South Australia

There are four classes of information system that are not well served by current modeling techniques. First, there are systems for which the number of instances for each entity is relatively low resulting in data definition taking a disproportionate amount of effort. Second, there are systems where the storage of data and the retrieval of information must take priority over the full definition of a schema describing that data. Third, there are those that undergo regular structural change and are thus subject to information loss as a result of changes to the schema's information capacity. Finally, there are those systems where the structure of the information is only partially known or for which there are multiple, perhaps contradictory, competing hypotheses as to the underlying structure.
This paper presents the Low Instance-to-Entity Ratio (LItER) Model, which attempts to circumvent some of the problems encountered by these types of application. The two-part LItER modeling process possesses an overarching architecture which provides hypothesis, knowledge base and ontology support together with a common conceptual schema. This allows data to be stored immediately and for a more refined conceptual schema to be developed later. It also facilitates later translation to EER, ORM and UML models and the use of (a form of) SQL. Moreover, an additional benefit of the model is that it provides a partial solution to a number of outstanding issues in current conceptual modeling systems.

4   Conclusion

We have summarized 11 regular papers and 3 invited papers for the 2006 First International Workshop on Active Conceptual Modeling of Learning. These papers represent some of the most current ideas and research results on this very important subject.

Architecture for Active Conceptual Modeling of Learning

T.C. Ting[1], Peter P. Chen[2,*], and Leah Wong[3,**]

[1] Computer Science & Engineering Department
University of Connecticut
371 Fairfield Road, Unit 2155
Storrs, CT 06269-2155, U.S.A.

[2] Computer Science Department, Louisiana State University
Baton Rough, LA 70803, U.S.A.

[3] Space and Naval Warfare Systems Center San Diego
Code 246214, 53560 Hull Street
San Diego, CA 92152, U.S.A.

ting@engr.uconn.edu, pchen@lsu.edu, leah.wong@navy.mil

Abstract. The concept of Active Conceptual Modeling of Learning (ACM-L) has been explored in order to capture content and context changes that permit a comprehensive learning from the past, understanding the present, and forecasting the future. Such capability has not been fully explored and it is not available with today's static oriented database system. The potential of creating a "database of intention" that can have its own aim to understand the intentions of its users and the changes to their environment. This paper explores an architectural approach for the "database of intention" with predictability power. The proposed architecture is presented, illustrated, and discussed.

1 Introduction

Human beings use lessons learned from past successes and failures in order to make better future decisions. The idea of learning from the past for future actions is not new, but the comprehensive collection and understanding of dynamic data and their environment has not been fully explored. The concept of "Active Conceptual Modeling of Learning" (ACM-L) has been explored for effectively capturing the dynamic data and changes in their environment so that it permits to learn comprehensively from them in order to understand the present, and predict the future [1-6].

---

* The research of this author was partially supported by National Science Foundation grant: NSF-IIS-0326387 and Air Force Office of Scientific Research (AFOSR) grant: FA9550-05-1-0454.
** Contributions to this article by co-author Leah Wong constitute an original work prepared by an officer or employee of the United States Government as part of that person's official duties. No copyright applies to these contributions.

P.P. Chen and L.Y. Wong (Eds.): ACM-L 2006, LNCS 4512, pp. 7–16, 2007.
© Springer-Verlag Berlin Heidelberg 2007

Having a clear understanding of the changing real world state is essential for intelligent decision making. Decision makers must be able to detect critical factors and events in the evolving domain of the real world and to learn from them in order to better plan for and handle the future. This intellectual capability can be augmented by a "database of intention" that can help in collecting and analyzing past events in the real world environment and in assisting human users for their decision making. Recent surprising incidents (e.g. September-11, Tsunami, Katrina, etc.) have motivated us to examine and reexamine what facts, activities, and trends that led to past incidents in order to better prepare for and manage similar events in the future. Predictive, preventive, and reactive strategies could be developed for better managing reoccurrences of similar incidents. When we review events, we turn back the clock relative to the timeframe and relate past information of events and facts with multiple perspectives, orientations, and view points to help clarify what we didn't and/or failed to see before. This "Enhanced Rewind" Paradigm within the framework of Conceptual Modeling of Learning enhances our intellective capability by reviewing the history of past events and their newly-discovered relationships, so that we are better equipped to learn lessons that will affect our responses to similar mishaps in the future.

1.1 The Challenge

The current information processing technologies can support only "simple rewind." For example, it is possible to use computer backup files to a certain date in the past, and it is also possible to retrieve a previous day's newspaper contents, etc. However, the detailed changes leading to most recent contents were often omitted. Therefore, the existing database and information system technology do not support the learning from the past very well.

The challenge is the development of next generation of information system that it will keep and provide detailed changes on records and their corresponding real world environment under which the changes were made. We are proposing to take the features of Active Conceptual Modeling that could capture and represent both the static and dynamic aspects of the real world domain. The persistently stored changes would be organized and interconnected for permitting effective learning from the collection. The traces of past data could be done with multiple dimensions and perspectives so that questions on "who, what, where, when and how" changes occurred could be answered. Significant sequences of events could be extracted from the comprehensive collection for intense review, analysis, and compare so that potentially hidden factors and implicit relationships could be uncovered.

1.2 An Enhanced Rewind Paradigm for Learning

An Enhanced Rewind Paradigm for Learning as one of the approaches in response to the challenge is identified and illustrated. The required features of the paradigm are as follows:

1. A continual process of modeling events and their changes within a given domain.
2. Backtracking of the events to the point and time of interest.
3. Identification of related and parallel events during the period when an incident occurred for possible hidden facts and relationships.
4. Modeling and learning of historical information from different perspectives.
5. Generating and triggering alerts on potentially incidents based on past experiences.

Active Conceptual Model of Learning is incorporated into the Paradigm to ensure the presentation of the real world dynamic environment. Multi-level and multi-perspective learning strategies are integrated and included for examining the dynamic collection with different learning logic and approaches. The goals and objectives of planners and decision makers and environmental constraints can be entered into the paradigm for providing guidance and focus of the learning and monitoring processes.

The following diagram depicts the Enhanced Rewind Paradigm for Learning.

Vision

1.3 The Concept of "Database of Intention"

Based on the proposed paradigm, a "database of intention" could be created based on the proposed paradigm for collecting and storing past actions and changes in environments, and learning from them. Such a database system focuses on the prediction of future intentions on users and their environmental changes based on their past behaviors. It augments human decision makers for taking planned actions. The system continuously monitors on going changes and continuously learning from the conceptually modeled dynamic real world environment for providing hints and alerts to human decision makers and planners. Specific constraints and conditions could be entered by human decision maker so that the system may learn from the past activities within the set of constraints provided by the human decision makers. The system is mainly for augmenting the human decision maker's intellectual capability for decision making under constraints. The "database of intention" expands the current database system into a new dimension for handling both static as well as dynamic data and information.

Based on the concept of "database of intension", a high level architecture for the next generation of database information technology is proposed. In this paper, the key technologies and components of the architecture are identified, illustrated, and discussed.

The proposed system architecture attempts to show the technical viability for the next generation of data and information system for providing a framework for the research and development community for the development of prototypes, experimental test beds, and eventually the full operational system in the not too distant future.

2 Proposed Architecture

System architecture is proposed to provide a framework for the development of the next generation of information system. The proposed system architecture is based on the core technology of an interconnected mass persistent storage that can store and organize massive changes data within the database and its domain data sources from the Internet and the real world. This mass persistent storage is managed by an executable conceptual model database management system so that it insures the consistency of data collection and data usage. The executable conceptual model database provides the flexibility for dynamically modeling the changing world by evolutionally changing the conceptual model as well as its associated data storage structures accordingly. The evolutional changes of the conceptual model are maintained in the persistent storage with their associated data to ensure that the changes in entity behavior as well as the traces and links that reveal "who, what, when, where and how" are directly associate with the correct version of the conceptual model in time.

The proposed architecture has three major interacting operational modules, namely, Data Acquisition Module, Learning Module, and Planning and Decision Making Module.

The following diagram depicts the proposed architecture.

System Architecture

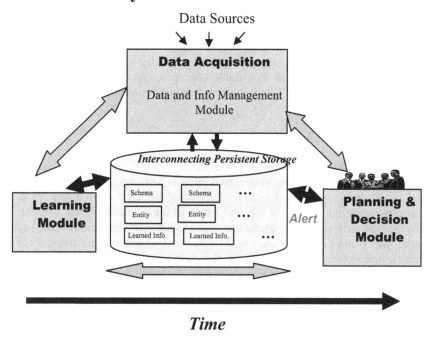

2.1 Executable Conceptual Model Database Management System

The executable conceptual model data management system works more like an interpreter that directly interprets the actions specified in the conceptual model. It is different from the conventional database management system which uses the layered architecture for accomplishing data independence and the conceptual model is used as a database design tool for guiding the generation of different levels of schema. The executable conceptual model database management system is intended to execute the database management function directly through the conceptual model. The data collected in the interconnecting persistent storage can be linked directly with its corresponding elements in the conceptual model. This is necessary in order to make clear traces from applications through processes to entities and their instances. With the executable conceptual model database management system the instances of the entity are directly traceable to the entity and the entity is directly traceable to the conceptual model and applications. Therefore, the data in the persistent storage are closely coupled with the conceptual model. In a conventional database management system, the conceptual model is stored in database design or CASE tools, the physical data in the database management system is isolated from the conceptual model and the application programs. Therefore, the architecture of conventional database management system created the difficulty for tracing storage data element back to its corresponding entity in the conceptual model. Reverse engineering has been used to accomplish this task without much success. The conventional architecture also introduced the difficult for evolutional changes in conceptual schema. When necessary, the database must be reorganized and regenerated which is a costly and time consuming task. Another important function of the active conceptual model is that it requires the handling of dynamic changes of the change of conceptual model itself in an evolutional manner. Schema evolution in a conventional database management system is a major difficulty that it requires the reorganization of the database with each modification of the conceptual model. By linking and integrating the major components together, the executable conceptual model data management system accomplishes traceability between application, conceptual schema, and storage structure.

2.2 The Mass Interconnecting Persistent Storage

The mass interconnecting persistent storage is organized by the conceptual model that represents the real world domain. This stored data is also operated by the same conceptual model that created the data store. Each database has a unified conceptual model that represents the domain of the real world of applications. When a data is changed in the persistent storage a link will be created to specify the current value and to link to its pass value. A time stamp will be inserted to indicate the time of the change with a link to the process that caused the change and the application and the user that activated the process. These links provide the trace to "what, when, how, where, and who" about the change.

The interconnecting persistent storage is designed to provide the sophisticated learning process to later analysis of changes. The interconnected persistent storage provides the needed links for comprehensive, multi-perspective, and multi-level learning to discover insights and clues from the past.

2.3 Three Interrelated Modules

The three interrelated modules each has a different mission governed by the executable conceptual model database management system. The data acquisition module identifies the data resources and captures the changes during the operations of the database. The learning module is the main learning engine for the analysis of the historical data. It may include multiple learning strategies and cognitive processes. The selected learning processes will be used during the conceptual modeling process so that characteristics of learning strategies can be incorporated into the conceptual model to be developed. Some of the learning processes that do not depend upon the structured data can also be included in the learning module but not used in conceptual modeling process. These learning processes, such as data mining, can be used to access the database, but not be governed by its structures and semantics by the conceptual modeling process. The planning and decision making module interface with the human decision makers for providing the decision makers with insights, clues, and alerts. Tools may be provided for validating results from different learning processes for their relevancy and accuracy and also allow decision makers to ask "what if" questions as well as focused questions about patterns concerning "who, what, where, when and how" changes occurred.

Data Acquisition Module. The data modeling and acquisition module controls the collection of dynamic data for storage in the interconnecting persistent storage and establishes the proper time stamps and links for every content and context changes. The collected active data in the interconnecting persistent storage are closely coupled with the conceptual model used so that their semantics would be consistent. Entities can be accessed only via the designated processes as indicated in the conceptual model so that the semantics of the accesses and data transformation are consistent as designed the conceptual modeling process. Data accesses that do not follow the pre-designed access patterns will have to be considered and recorded as unspecified accesses. Unspecified accesses are not traceable and therefore, the data sources could not be checked and verified. Access integrity is extremely important to ensure that the active data collected for later learning can have high reliability and accuracy. The key element of the data acquisition is the indices and links that created when a change is occurred. Precedent and antecedent relationships of changes allow the building of indices that can reveal "when, what, where, who, and how" the change occurred, facilitating "enhanced rewind" for later learning processes. The comprehensive set of indicators, links, and pointers are the core of the architecture for active conceptual modeling. The fiber of this interconnecting storage is the main feature for supporting the comprehensive learning and inference strategies which is the hallmark of the active conceptual modeling for the next generation information technology.

Learning Module. Learning is the key for the active conceptual model. There is a wide range of approaches in learning or cognition from ad hoc data mining that based on stochastic or chaos theory to well formulated time series that based on straight forward computation or observation. Different learning strategies offer certain advantages as well as having certain weaknesses.

The Inclusion of Multiple Learning Strategies. The learning module allows the inclusion of different learning strategies and approaches for accessing the interconnecting persistent storage. Enhanced rewind feature permits the learning from the past by following the time dimension or tracing a set of specially selected features from the interconnecting persistent storage.

Learning strategies that follow the structures and semantics of the conceptual model could produce reliable and interpretable results, but they may miss potential cues that were not expressed in the conceptual model. Other learning approaches such as data mining which treats the collection of data as an unstructured repository of data, may find relationships between various data elements that were not anticipated by the conceptual modeler. These seemly unrelated relationships may provide discoveries beyond the anticipation of the data model designer. However, the validity and relevancy of these discoveries need to be carefully checked, assessed, and interpreted. In order to have comprehensive learning capability learning strategies with different strategies of both types are included.

The Comprehensive Learning Engine. Users can be augmented by using various learning capabilities provided by different learning strategies and processes. The inclusion of multiple learning strategies can be used collectively and intelligently as a comprehensive learning engine. Specially demanded focused learning is feasible. Learning can also be pre-programmed for automatic and/or continuously learning over the dynamically expanded period of data or follow specially identified target with special features and critical factors.

The learned information may be used to establish certain norms or lesson learned from the past. These norms or lessons learned could be used to monitor on going situations for generating alters, clues when the situation is sufficiently deviated from the norm or certain indications that the lessons learned in the past could be repeated soon.

Planning and Decision Making Module. Planning and Decision Making Module interfaces with human users, augmenting their intellectual ability in planning and decision making tasks. The conceptual model provides navigational map for accessing the persistent storage and for guiding the interpretation of the results. "What if" questions may be asked, explore potential situations under different condition and assumption.

Automatic Alert subsystem. Using the lesson learned or other pre-established norms or patterns, the ongoing activities can be monitored. Human decision makers, analysts, or planners could be automatically altered when the potential abnormal situation is detected. The learning module can generate information for the establishment of norms and warning signals. When the current operation deviates sufficiently from the established norm warning signal would be generated.

Demanded Learning with specific conditions and assumptions. Human users submit request with a set of conditions and/or assumptions for learning from a selected set of data. This feature supports the function of asking "what if" questions. The collaboration and interaction between the human user and the system is done with the purpose of augmenting the human user's intellectual capability for planning and decision making under constraints.

2.4 Other Important Features and Capabilities

Besides the major modules mentioned earlier, we think a system to be developed based on this architecture should have at least several other important features:

- Information Provenance [7-10]: The system should be able to keep track of the pedigree of the data/information. In particular, the system should follow the 7W model [10] and be able to keep track of and answer queries of the questions on "Who", "When", "What", "Where", "How", "Which", and "Why".
- Visual Query Interface [11]: The Entity-Relationship Diagram or similar graphical representation will be a very good user interface for user to interact with the system.
- Automatic or Semi-automatic handling of schema and conceptual model changes [12]: It is very difficult and time-consuming for existing systems to change schemas and/or conceptual models. It will be a welcome feature to provide the capability to assist this change process.

3 Impact

We make decisions based, in part, on our understanding of what has already happened. The proposed system architecture for active conceptual model of learning addresses a number of existing challenges:

- Information from multiple modalities and sources needs to be combined and synthesized in ways that are general and replicable.
- Analysts require the discovery of hidden and implicit relationships between different events.
- Planners need to trigger alerts and to plan combat or non-combat operations based on lessons learned from similar past events.
- Decision makers need to understand and learn lessons from the past to make better decisions for the future.
- Decision makers need to make decisions with greater speed and insight as conditions change, and more effectively handle complex operational events.

4 Possible Applications and Conclusion

Using this architecture framework, we may imagine what tools can be available in the near future for the analysts, planners, and decision makers to make better analyses, plans, and decisions. The following are two military applications::

(1) "Enhanced Rewind" which identifies the relevant events that led to the planting of an **improvised explosive device** (IED). From the source of the IED material, assembly place of the IED, transportation methods of IED, and hideout locations of the insurgents, we can make attack plans and predict the future locations of IEDs from the same group of insurgents.

(2) "Enhanced triggers" which will alert the military planners in the Pacific Disaster Center (located on Maui) preparing for Evacuation Operations (EO) for the predicted forthcoming natural disasters (Tsunami, etc.) or to alert law enforcement agencies or Homeland Security Personnel of potential terrorists [13].

We believe the proposed architecture will provide a framework and guidelines in the development of the next generation of information system to make such scenarios feasible. The architecture provides a foundation for building the next generation of information technology that can manage dynamic data.

References

1. Vego, M.N.: Effects-Based Operations: A Critique. Joint Forces Quarterly (41) (2006)
2. Chen, P.P.: Entity-Relationship Model: Towards a Unified View of Data. ACM Transactions on Database Systems 1(1), 9–36 (1976)
3. ER Conference website, http://www.conceptualmodeling.org

4. Chen, P.P., Thalheim, B., Wong, L.Y.: Future Directions of Conceptual Modeling. In: Chen, P.P., Akoka, J., Kangassalu, H., Thalheim, B. (eds.) Conceptual Modeling. LNCS, vol. 1565, pp. 287–301. Springer, Heidelberg (1999)
5. Chen, P.P., Wong, L.Y.: A Proposed Preliminary Framework for Conceptual Modeling of Learning from Surprises. In: Proceedings of the International Conference on Artificial Intelligence (ICAI) 2005, pp. 905–910. CSREA Press (2005)
6. Chen, P.P.: Suggested Research Directions for a New Frontier - Active Conceptual Modeling. In: Embley, D.W., Olivé, A., Ram, S. (eds.) ER 2006. LNCS, vol. 4215, pp. 1–4. Springer, Heidelberg (2006)
7. Dublin Core Collection Description Working Group, Proposed Definition of "Provenance" (March 14, 2004), http://www.ukoln.ac.uk/metadata/dcmi/collection-provenance/
8. Rogerson, S.: Information Provenance. ETHI column in the IMIS Journal 15(1) (2005), http://www.ccsr.cse.dmu.ac.uk/resources/general/ethicol/Ecv15no1.html
9. UIUC Department of Urban and Regional Planning, The LEAD Data Pedigree System, http://wiki.cs.uiuc.edu/ProjectIdeas/LEAM+Data+Pedigree+System
10. Ram, S., Liu, J.: Understanding the Semantics of Data Provenance to Support Active Conceptual Modeling. In: Proceedings of the 1st International Workshop on Active Conceptual Modeling of Learning. LNCS, vol. 4512, Springer, Heidelberg (2007)
11. Thalheim, B.: Visual SQL: an ER-Based Introduction to Database Programming. Research Project, University Kiel, Germany, working paper
12. Luqi, L.D.: Schema Changes and Historical Information in Conceptual Models of Learning, working paper (2004)
13. Chen, P. P. (principal investigator): Research project on Profiling Methods for Anti-terrorism and Malicious Cyber Transactions, NSF Grant #IIS-0326387, 8/03 - 7/08

Understanding the Semantics of Data Provenance to Support Active Conceptual Modeling

Sudha Ram and Jun Liu

430J McClelland Hall, Department of MIS, Eller School of Management
University of Arizona, Tucson, AZ 85721
{ram,junl}@eller.arizona.edu
http://adrg.eller.arizona.edu/

Abstract. Data Provenance refers to the lineage of data including its origin, key events that occur over the course of its lifecycle, and other details associated with data creation, processing, and archiving. We believe that tracking provenance enables users to share, discover, and reuse the data, thus streamlining collaborative activities, reducing the possibility of repeating dead ends, and facilitating learning. It also provides a mechanism to transition from static to active conceptual modeling. The primary goal of our research is to investigate the semantics or meaning of data provenance. We describe the W7 model that represents different components of provenance and their relationships to each other. We conceptualize provenance as a combination of seven interconnected elements including "what", "when", "where", "how", "who", "which" and "why". Each of these components may be used to track events that affect data during its lifetime. A homeland security example illustrates how current conceptual models can be extended to embed provenance.

1 Introduction

Data Provenance refers to the lineage or history of information including its origin, key events that occur over the course of its lifecycle, and other details associated with its creation, processing, and archiving. It is the background knowledge that enables a piece of data to be interpreted correctly and to support learning. We believe that tracking provenance, such as the processing and usage history of data, enables users to share, discover, and reuse the data, thus streamlining collaborative activities and reducing the possibility of repeating dead ends.

Despite its critical importance, current approaches to capturing provenance of data have not been particularly effective. As suggested by [1], data provenance needs to be captured with the hope that it is comprehensive enough to be useful in the future. However, due to the lack of consensus on the semantics or meaning of provenance, the concept has not been well-defined in the literature. For instance, some researchers define provenance as the origin of data and its movement between databases [2], while others view it as the process of transformation of data [3]. Accordingly, current

P.P. Chen and L.Y. Wong (Eds.): ACM-L 2006, LNCS 4512, pp. 17–29, 2007.
© Springer-Verlag Berlin Heidelberg 2007

efforts aimed at capturing data provenance typically focus on some aspects of provenance while ignoring others. As an example, [4] identifies two kinds of provenance – "why" and "where". The former refers to the source data that had some influence on the creation of the data of interest; the latter specifies the location(s) in the databases from which the data was extracted. We believe that provenance includes more than what is captured in [4]. In some application domains, provenance may include the literature reference where data were first reported, the history in terms of how the data was created and transformed, the series of experimental procedures by which it was derived from other data, and the sequence of ideas leading to an experiment. Consequently, to generate a complete record of data provenance, it is desirable to gain a deep understanding of the semantics of provenance and identify the key concepts associated with it. To our knowledge, none of the existing work has explored the "semantics" of provenance.

The primary goal of our research is to investigate the semantics or meaning of data provenance. We have developed a generic model called the W7 model that represents data provenance as a combination of seven interconnected elements including, "what", "when", "where", "how", "who", "which", and "why". Each of these elements may be used to track provenance and may be applied to different domains such as homeland security. Further, we demonstrate how our W7 model can help in active conceptual modeling.

2   Provenance Semantics – The W7 Model

Conventional conceptual models do not provide a straightforward mechanism to explicitly capture the semantics of data provenance, and it is still unclear how provenance information can be linked with the application data at the conceptual level. In response to this problem, we propose an ontological model called the W7 model to capture the semantics of data provenance.

2.1   Theoretic Basis – Bunge's Ontology and Other Philosophical Work

To understand the semantics of provenance and identify concepts associated with it, we use Bunge's ontology [5, 6] as our starting point. Table 1 summarizes the concepts from Bunge's ontology that are appropriate for understanding data provenance.

The data stored in an information system is a thing and thus has history that is represented as a sequence of events or state changes. Based on the observation that data provenance describes history of data, we propose to define provenance by recording all events that happen to data during its lifetime. Each of these events happens to a data object when it is created or destructed or when it is acquires/loses one or more of its properties or changes the values of its properties. In database applications, these events center around the lifecycle of data which includes creation, updates, and deletion of data.

However, simply recording *what* events occur is not sufficient to meaningfully represent the provenance of data. To provide insightful provenance knowledge, it is necessary to identify and explicitly describe various details describing the events. Bunge's ontology includes some constructs related to events such as space, time and actions. According to [5], space and time provide the basic framework of events. An event happens and history of a thing changes when it is under actions of an agent or other things. Hence, a causal relationship exists between an action and an event [5]. A further investigation of the relationship between events and actions leads us to other philosophical work such as [8] and [9] that help identify other important concepts related to events. In [8], Davis defines an action as a "doing" in which an event occurs when an agent wants or has other reasons for the event to happen. An action provides the causal explanation, i.e., "how"-explanation, while the agent's intentions, beliefs and other mental events provide the reason explanation, i.e., "why"-explanation for the event. Chisholm, on the hand, stresses the importance of the agent for an event by proposing the notion of "agency causation": An agent brings about an event by undertaking something [9]. Based on these philosophical works pertinent to events and actions, we identify concepts that provide insights to events including actions (how), reasons (why) and agents (who and which) in addition to space (where) and time (when).

Table 1. Selected Ontological concepts from Bunge (adapted from [5]and [7])

Ontological Concept	Explanation
Thing	A thing, the elementary unit in the ontology, is a substantial individual endowed with its properties. Two or more things (composite or simple) can be associated into a composite thing
Property Intrinsic Mutual	Things possess properties. A property that is inherently a property of a thing is called an intrinsic property. A property that is meaningful in the context of two or more things is called a mutual or relational property
State	The property value vector of a thing at a given time
Event Qualitative Quantitative	An event is a change of state of a thing. An event of a thing is qualitative when it acquires or loses one or more properties. An event is quantitative when the thing changes its property values
History	A sequence of events or state changes of a thing
Action	The history of a thing changes when it is under actions of other things
Space and time	Everything exists in space and time. The history of any thing has a nonempty projection onto the spacetime, a unification of space and time

2.2 Overview of the W7 Model

Based on our theoretical analysis of data provenance, we conceptualize provenance as consisting of seven interconnected dimensions including what, when, where, who, how, which, and why.

Definition 1. Provenance is defined as a n-tuple P = (WHAT, WHEN, WHERE, HOW, WHO, WHICH, WHY, OCCURS_AT, HAPPENS_IN, LEADS_TO, BRINGS_ABOUT, IS_USED_IN, IS_BECAUSE_OF), where

- *WHAT* denotes the sequence of events that affect the data object; *WHEN*, the set of event time; *WHERE*, the set of all locations; *HOW*, the set of all actions leading up to the events; *WHO*, the set of all agents involved in the events; *WHICH*, the set of all devices; *WHY*, the set of reasons for the events. The formal definition of each of these 7 Ws is given later.
- *OCCURS_AT* is a collection of pairs of the form *(e, t)*, where $e \in$ *WHAT* and t $\in$ *WHEN*. *HAPPENS_IN* is a collection of pairs of the form *(e, l)*, where l denotes a location. *LEADS_TO* represents pairs of the form *(e, h)*, where $h \in$ *HOW* and represents an action leading up to the event *e*. *BRINGS_ABOUT* is a collection of pairs *(e, {a₁, a₂, ..., aₖ})*, where $a_1, a_2, ..., a_k \in$ *WHO*, indicating that more than one agent can cooperate to bring about an event, and *IS_USED_IN* is also a collection of pairs *(e, {d₁, d₂, ..., dₖ})*, where $d_1, d_2, ..., d_k \in$ *WHICH*. Finally, *IS_BECAUSE_OF* is a collection of pairs of the form *(e, {y₁, y₂, ..., yₖ})*, where y_1, $y_2, ..., y_k \in$ *WHY*.

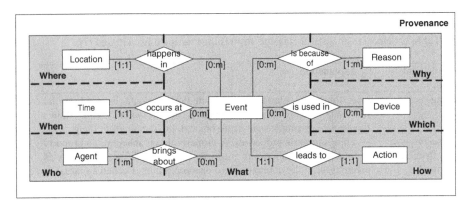

Fig. 1. Overview of W7 Model

Fig. 1 provides an overview of the W7 model. We allow data provenance to be specified for data objects at different granularity levels. For example, for some applications, provenance may be specified on instances of an entity class, and their provenance be aggregated to form the provenance of the entity class or a subset of it. Alternatively, data objects that are subset of a parent dataset may share their provenance with the parent and yet be different as a whole. "What", i.e., a sequence of events, is the anchor of our W7 model. The other provenance components are semantically related to "what" and describe various details about the events. An event "occurs at" a time and "happens in" a location. A "leads_to" relationship exists between "action" and "event", indicating the causal relationship between actions and events. An action is taken by some *agents* using some *devices* for some *reasons*, which is reflected by the various relationships existing between "what" and the elements "who", "which" and "why". We discuss the semantics of each element and provide a graphical representation using the Unifying Semantic Model (USM) [10], an extended version of the Entity-Relationship (ER) model [11]. The USM is used in provenance modeling for two purposes. First, the USM extends the ER model with constructs such as super/subclasses (represented by "⎯ⓔ► ") and groups/aggregates (represented by " ◄Group► "), thus providing a formal and precise expression of provenance semantics. Second, it enables us to directly apply the W7 model to conceptual modeling with easy adaptations.

2.3 What

The fundamental building block of the W7 model is the element "what". Its semantics are defined as follows.

Definition 2. *WHAT* is a sequence of events $<e_1, e_2, ..., e_n>$ that affect a data object during its life time.

We categorize events by drawing upon Bunge's theory. A data object such as an entity instance in a relational database or a document in a digital library is a thing often composed of other things (e.g. an entity instance is a composite thing made up of attribute values). A data object as a thing has intrinsic and mutual or relational properties. The intrinsic properties of a data object normally include its content and composition, while its mutual properties of a data object provide its contextual information such as its ownership, custody, rights, etc. An event of thing is a change of state. In addition to creation or "coming into being" event [6] and destruction, an event happens to a data object when it acquires or loses one or more of its intrinsic or mutual properties (i.e., qualitative events) or when it changes its property values (i.e., quantitative events). Accordingly, we classify events into creation, transformations, destruction, and contextual events. Transformations of a data object represent changes that happened to its intrinsic properties such as its content. Contextual events are manifested by changes made to mutual properties. They can be further classified into ownership changes, changes of storage location, etc. Fig. 2 presents a graphic representation of "what".

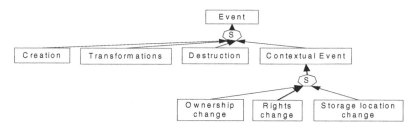

Fig. 2. Semantics of "What"

2.4 When

The semantics of "when" are shown in Fig. 3. Different from existing temporal data models such as [12, 13] that capture valid time, i.e., a time period during which a fact is true in the real world, and transaction time, i.e., a time period during which a fact is stored in the database, our model focuses on recording time of various events that affect data during its lifetime. As an example, given a script of a correspondence between two terrorist suspects, we capture the duration or time period during which the script is recorded as its creation time. When the script is stored in a database, we record the data storage time. When the script is accessed/used, we capture the time period during which it is used. Associating a timestamp with each event provides a detailed timeline of the events and enables us to reconstruct the history of the data.

Definition 3. *WHEN* represents a set of timestamps $\{t_1, t_2, ..., t_n\}$ associated with various provenance events.

While some events may be instantaneous, others may occur over an interval of time. Accordingly, we specify two disjoint subsets of *WHEN* including *INTANT* and *DURATION*. *INSTANT* is a set of instants. Each instance is a point on the time line. *DURATION* represents a set of durations. A duration refers to the time between two instants with a start and an end. Hence, let *INSTANT_VALUE* denote a set of values an instant can take, and we define two functions: *Start: DURATION* $\rightarrow$ *INSTANT_VALUE* and *End: DURATION* $\rightarrow$ *INSTANT_VALUE*. Moreover, *DURATION* should be well-formed, which entails a constraint $\forall t \in$ *DURATION*, *Start(t)<End(t)*.

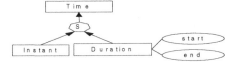

Fig. 3. Semantics of "When"

2.5 Where

The element "where" in the W7 model captures event locations. We provide a graphic representation of "where" in Fig. 4.

Definition 4. *WHERE* denotes a set of locations $\{l_1, l_2,..., l_n\}$, where various events happen.

The most common forms of representing locations are physical and geographical. Physical locations specify the position of places or points based on a global coordinate system, while geographical locations signify an area or boundary governed by a common law and are normally organized hierarchically. Correspondingly, we capture these two concepts as two subsets of *WHERE* including *PHYSICAL_ LOCATION* and *GEOGRAPHICAL_LOCATION*. In addition to physical and geographical location, we introduce the concept of *logical location*, which links a data object to its location in a server or database. This concept is important since data may travel between information sources due to location change events such as storage and transfer. The logical location can often be represented by a URI.

A location can be typed into *source* and *destination*. Let *LOCATION* represent a set of locations, and we specify a function that maps a transaction location to its type as *Type: LOCATION* $\rightarrow T$, where $T = \{Source, Destination\}$.

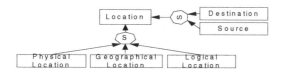

Fig. 4. Semantics of "Where"

2.6 How

"How" documents *actions* that lead to the occurrence of an event. An action is seen as a system of "doings", where agents work on certain objects in order to obtain a desired outcome. Actions are causes of event, and events are brought into being as results of actions performed by agents. Consider the case of creation of a data object. A data object can be created as a result of an action such as an observation or a measurement. It can also be generated by deriving it from existing data or acquiring it from other people.

Information regarding actions normally includes:
- *Preconditions* that refer to conditions that must hold prior to the enactment of an action.
- *Methods* that provide detailed descriptions about what has been done and capture various action parameters.
- *Inputs* and *outputs* that refer to data objects that are manipulated by the enactment of an action. An action can thus be seen primarily as a process of transforming a set of inputs into outputs.
- *Sources* refer to people or media we acquire data from. As an example, a CIA agent is a source of a terrorist report.

Definition 5. *HOW* is defined as a tuple (ACTION, PRECONDITIONS, METHODS, INPUTS, OUTPUTS, SOURCES, $\mathcal{P}, \mathcal{M}, I, O, \mathcal{S}$), where

- *ACTION* = $\{h_1, h_2, ...,h_n\}$ is a set of actions, and the previously mentioned concepts such as *Preconditions, Methods, Inputs,* and *Resources* are also defined as sets.
- $\mathcal{P}$: *ACTION* → *PRECONDITIONS* is a function that maps an action to its preconditions; $\mathcal{M}$: *ACTION* → *METHODS* maps an action to its methods; *I:* *ACTION* → *INPUTS* maps an action to its inputs; *O: ACTION* → *OUTPUTS* maps an action to its outputs; and *S: ACTION* → *SOURCES* associates an action with the sources.

Following [14], we classify actions into *primitive* and *complex* (see Fig. 5). Accordingly, we specify that *ACTION* consists of two subsets *PRIMITIVE_ACTION* and *COMPLEX_ACTION*. An action is considered to be primitive if no decomposition will reveal any further information which is of interest. Complex actions, on the other hand, may be arbitrarily complex activities and can be decomposed into primitive actions that happen sequentially or simultaneously. Moreover, previous research such as [3] has been focused on capturing the procedures used for processing the data, by describing the workflow of an experiment.

Accordingly, we define a "depends_on" relationship that captures the control flow of primitive actions within a complex action such as concurrency, sequence, etc.

Definition 6. A complex action c = (P, DEPENDS_ON), where

- $P = \{p_1, p_2, ..., p_k\}$ is a set of primitive actions that constitute the complex action c.
- $DEPENDS\_ON = \{d_1, d_2,..., d_k\}$ is a set of relationships. Each relationship d is an ordered pair (p_i, p_j), where $p_i, p_j \in P$.

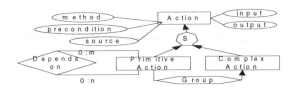

Fig. 5. Semantics of "How"

2.7 Who

"Who" refers to *agents* who bring about the events. The USM diagram of "who" is shown in Fig. 6. An agent is "an intentional entity", that is it has some idea of purpose that guides its actions [14]. We use the term "agent" refer to human agent including *persons* and *organizations*. Artificial agents such as software applications are captured in "which". An agent assumes a *role* in an event, and a role is defined as "a coherent set of activities to be assigned to an agent as a functional responsibility"[15]. Each agent plays a certain role to make some contributions to the action leading up to an event. For instance, a federal agent may play the role of supervisor in creating the script of a suspicious correspondence.

Definition 7. *WHO* is a triple (AGENT, ROLE, $\mathcal{RL}$), where

- $AGENT = \{a_1, a_2, ...,a_n\}$ is a set of agents that are involved in various events.
- $ROLE= \{r_1, r_2, ...,r_n\}$ is a set of roles agents are allowed to assume.
- $\mathcal{RL}$: $AGENT \rightarrow ROLE$ is a function that associates an agent with the role she played in a particular event.

WHO includes two subsets, i.e., a set of persons *PERSON* and a set of organizations *ORGANIZATION*. We often need to capture the *position* and *affiliation* of a person. When a person participates in her affiliation, she is no longer entirely free to choose her goals and actions. Instead, she accomplishes some activities according to her position. A *position*, which is called organizational role in [14], represents a set of responsibilities of an individual in her affiliation. As a result, we specify a function $\mathcal{PA}$: *PERSON* $\rightarrow$ *POSITION* $\times$ *AFFILIATION* that maps a person to her position and affiliation.

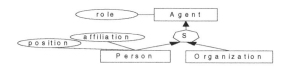

Fig. 6. Semantics of "Who"

2.8 Which

The element "which" describes which *devices* are used in data creation, analysis, and transformation. Devices can be distinguished into *instruments* (e.g. equipments and hardware) and *applications*. When an event involves a device, some level of detail about the device in which it is hosted should be captured. Moreover, some actions are specifically supported or offered by certain devices, whereby the characteristics and capability of the devices may play an integral role in describing the behavior of the action.

As shown in Fig. 7, the information related to a device is logically divided into three classes depending on the type of information they provide, namely device *description*, *function* and *settings*. Device description contains basic information related to a device such as its name, vendor, version, etc. A device's function can be specified in terms of the variables of the device itself, e.g., a battery's function is often specified as providing an electric voltage measured in volts. More frequently, a device is composed of parts or components, and its function is expressed in terms of the variables of its components. As an example, a computer may have a CPU of 2.0 GHz and memory of 256 MB. Different from the functional properties that rarely change throughout a device's lifetime, its *settings* contain volatile information pertaining to the device such as current level of CPU usage and remaining power level of a computer. The settings of a device often vary among applications, and it specifies the performance of the components of a device during an event.

Definition 8. *WHICH* is a tuple (DEVICE, SETTINGS, DESCRIPTION, FUNCTION, S, $\mathcal{D}$, $\mathcal{F}$), where

- *DEVICE* = $\{d_1, d_2, ..., d_n\}$ is a set of devices used in various events. It consists of two disjoint subsets *INSTRUMENT* and *APPLICATION*.
- *SETTINGS* denotes a set of settings a device can take, *FUNCTION* represents a set of device functions, and *DESCRIPTION* denotes a value set of descriptions a device can take.
- S: *DEVICE* → *SETTINGS*, $\mathcal{D}$: *DEVICE* → *DESCRIPTION*, and $\mathcal{F}$: *DEVICE* → *FUNCTION* represent mappings from a device to its settings, description, and function.

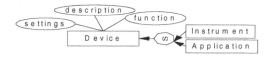

Fig. 7. Semantics of "Which"

2.9 Why

In this subsection, we define the semantics of "why" and provide a USM representation of the semantics in Fig. 8.

Definition 9. *WHY* represents a set of reasons $\{y_1, y_2, ..., y_n\}$ for various provenance events.

Our scheme for representing "why" is based largely on the Belief-Desire-Intention Model [16], which identifies *beliefs*, *desires* and *intentions* as significant factors that affects decision making. Beliefs represent knowledge of the world, desires are goals assigned to the agent, and intentions are commitments by an agent to achieve particular goals. Here, we collapse desires and intentions into *goals*. As a result, we specify two subsets of *WHY*, i.e., *BELIEF* and *GOAL*. The former represents a set of beliefs and the latter a set of goals.

A natural way to answer "why" questions is by tracing them to goals. For example, why a milestone is established in an anti-terrorist action can be related to the goal that the action be completed on time. Explicit representation of goals is important because it allows us to study a specific event from an intentional point of view. We also define an "is_reduced_to" relationship to capture the goal-subgoal structure (See Fig. 8). This relationship corresponds to the classical reduction operator in the problem reduction approach to problem solving. A goal can have several parent goals as it can occur in several reductions. We define $IS\_REDUCED\_TO = \{s_1, s_2, ..., s_n\}$ as a set of relationships representing goal-subgoal structures. Each relationship s is an ordered pair (g_i, g_j), where $g_i, g_j \in GOAL$. Furthermore, each relationship $s \in IS\_REDUCED\_TO$ should not be symmetric. Thus, we specify a constraint on s as $s \in IS\_REDUCED\_TO$ and $s^{-1} \in IS\_REDUCED\_TO \Rightarrow s = s^{-1}$

The other important concept associated with "why" is the concept of *belief*. Agents have a subjective view of the world, where they form their beliefs. Different from the goals an agent intends to fulfill through an action, beliefs refer to what an agent believes prior to the action, and they form the background upon which an agent can choose to act in a particular way [17]. We further classify beliefs into *assumptions* and *hypotheses* (see Fig. 8).

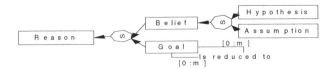

Fig. 8. Semantics of "Why"

3 Active Conceptual Modeling with Provenance Annotations

As discussed in [18], a serious problem in today's data modeling practices is that database design approaches have viewed data models as representing only a snapshot of the world and recommend ignoring variations of information as well as the causes and other details of those variations during data modeling. In response to this problem, Chen et al. propose active conceptual modeling [19] that describes all aspects of the world, its activities, and its changes under different perspectives, thus providing a multilevel and multi-perspective view of reality. Active conceptual modeling requires us to capture provenance knowledge in terms of *what* event/change may happen to the data, at the stage of conceptual modeling. Moreover, we need to identify provenance components such as "where", "when", "how", "who", and "why" behind the "what" to provide insights about the changes. Our W7 model captures the semantics of the various provenance components, thus providing a foundation for explicitly capturing data provenance in active conceptual modeling.

We propose to capture data provenance requirements of the users using provenance annotations. Rather than creating new constructs in a conceptual model, we augment a conventional conceptual model such as an Entity-Relationship model with annotations that represent the provenance information associated with data captured in the core data model without changing the core data model. Using an example in the domain of homeland security, we illustrate our annotation-based approach to active conceptual modeling with provenance annotations.

Provenance knowledge is indispensable in various applications. In particular, it is critical in the domain of homeland security, where given some intelligence information, provenance regarding the information such as how and when it was collected by whom is required to evaluate the quality of the information and avoid false intelligence. Consider the homeland security application described in the core data model in Fig. 9. Nowadays, organizations and ordinary citizens are called upon to report suspicious activities that might indicate terrorist threats. As a successful example, the Pan American Flight School reported that Zacarias Moussaoui seemed "extremely interested in the operation of the plane's doors and control panel", which leads to Moussaoui's subsequent arrest prior to 9/11 [20]. However, intelligence information such as terror reports may be false, out-of-date, and from unreliable sources, which calls for a data model with provenance semantics. Hence, we record events such as the creation and transformations of a terror report at the conceptual level using provenance annotations and by doing so, make the conceptual schema "active".

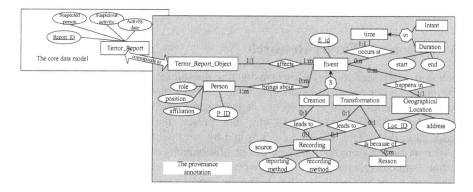

Fig. 9. A Sample Conceptual Model with Provenance Annotations

The annotation shown in Fig. 9 is developed based on our W7 model and is represented in the USM. The entity "Terror_Report_Object" in the annotation corresponds to the entity "Terror_Report" defined in the core data model and represents instances of "Terror_report" (see Fig.9). "Who", in this case, describes persons who play different roles in creating the data such as the agent who receives the report as well as the agent who records it. The actions leading up to the creation or transformations are normally recordings. We capture the source (i.e. the informant) of the report and instantiate the attribute "method" in the W7 model into "reporting method" and "recording method". When a transformation/update event happens to the data, we capture reasons "why" the information is updated in addition to "how", "who", "where" and "when". Via annotations, we enable a supplementary layer of abstraction that describes the data provenance semantics and naturally extends the semantics of a conventional data model.

Recording data provenance is critical in the domain of homeland security. It supports the following activities:

- *Information reliability*: To enforce national security, the right people must collect the right information from the right sources to identify real security threats. In our example, capturing who reported the threat via what reporting method assists in evaluating data reliability. Provenance regarding how the report was recorded or updated by who also helps ensure that the information can be trusted.
- *Information currency*: Some types of intelligence information may have a very short shelf-life. As an example, after Saddam Hussein fled Baghdad, information about him being spotted at a specific location changed six to eight times a day [20]. Capturing provenance such as: when the report of his being spotted was created and updated could be used to avoid being misled by old or out-of-date information.
- *Pattern recognition*: Provenance could help discover certain out-of-the-norm behavior patterns, which would helpful for predicting and preventing potential terrorist threats. As an example, a sudden increase in the number of threat reports from people in the same region within a short time period may indicate a terrorist plot. Also, the "who" part of our provenance could help us identify key reliable sources and forestall unreliable sources from feeding false intelligence.

4   Conclusion and Future Research

In conclusion, our research focus is on investigating the semantics of provenance. We have developed a generic provenance model, i.e., the W7 model, to represent these semantics. We identify various elements of provenance such as "what", "where", "when", "who", "how", "which" and "why" and present the semantics of each of these aspects in detail. Our W7 model is inspired by our investigations of theoretical works such as Bunge's ontology as well as our observations of provenance issues in application domains including biology, new product design and development, digital archiving and homeland security. It is a generic model of data provenance and is intended to be easily adaptable to represent domain or application specific provenance requirements in conceptual modeling. Using homeland security as an example application, we apply our W7 model to support active conceptual modeling with provenance annotations. We are investigating how provenance might be automatically identified and recorded. In the future, we will investigate the effectiveness of our approach by applying it to different application domains.

References

1. Pearson, D.: The Grid: Requirements for Establishing the Provenance of Derived Data. In: Workshop on Data Derivation and Provenance, Chicago, Illinois (2002)
2. Buneman, P., Khanna, S., Tan, W.C.: Data Provenance: Some Basic Issues. In: Kapoor, S., Prasad, S. (eds.) FST TCS 2000. LNCS, vol. 1974, Springer, Heidelberg (2000)
3. Frew, J., Bose, R.: Earth System Science Workbench: A Data Management Infrastructure for Earth Science Products. In: The 13th International Conference on Scientific and Statistical Database Management, Fairfax, VA (2001)
4. Buneman, P., Khanna, S., Tan, W.C.: Why and Where: A Characterization of Data Provenance. In: Van den Bussche, J., Vianu, V. (eds.) ICDT 2001. LNCS, vol. 1973, Springer, Heidelberg (2000)
5. Bunge, M.: Treatise on Basic Philosophy. In: Ontology I: The Furniture of the World, vol. 3, Reidel, Boston, MA (1977)
6. Bunge, M.: Treatise on Basic Philosophy. In: Ontology II: A World of Systems, vol. 4, Reidel, Boston, MA (1979)
7. Wand, Y., Weber, R.: On the deep structure of information systems. Information Systems Journal 5 (1995)
8. Davis, L.: Theory of action. Prentice-Hall, Englewood Cliffs, NJ (1979)
9. Chisholm, R.: The Agent as Cause. In: Brand, M., Walton, D. (eds.) Action Theory, D. Reidel, Dordrecht (1975)
10. Ram, S.: Intelligent Database Design Using the Unifying Semantic Model. Information and Management 29, 191–206 (1995)
11. Chen, P.P.: The entity-relationship model - toward a unified view of data. ACM Trans. Database Syst. 1, 9–36 (1976)
12. Snodgrass, R.T., Ahn, I.: Temporal Databases. Computer 19, 35–42 (1986)
13. Khatri, V., Ram, S., Snodgrass, R.: Augmenting a Conceptual Model with Geospatio-temporal Annotations. IEEE Trans. Knowledge and Data Eng. 16, 1324–1338 (2004)
14. Koubarakis, M., Plexousakis, D.: A formal framework for business process modeling and design. Information Systems 27, 299–319 (2002)
15. Curtis, B., Kellner, M., Over, J.: Process modeling. Communication of ACM 35, 75–90 (1992)
16. Georgeff, M., Pell, B., Pollack, M., Tambe, M., Wooldridge, M.: The Belief-Desire-Intention Model of Agency. In: The 5th International Workshop on Intelligent Agent: Agent Theories, Architectures, and Languages, Paris, France (1999)
17. Konolige, K., Pollack, M.E.: A Representationalist Theory of Intention. In: IJCAI 1993. The Thirteenth International Joint Conference on Artificial Intelligence (1993)
18. Allen, G., March, S.: Modeling Temporal Dynamics for Business Systems. Journal of Database Management 14, 21–36 (2003)
19. Chen, P.P., Thalheim, B., Wong, L.: Future direction of conceptual modeling. In: Chen, P.P., Akoka, J., Kangassalu, H., Thalheim, B. (eds.) Conceptual Modeling. LNCS, vol. 1565, pp. 294–308. Springer, Heidelberg (1999)
20. English, L.: Information Quality: Critical Ingredient for National Security. Journal of Database Management 16, 18–32 (2005)

Adaptive and Context-Aware Reconciliation of Reactive and Pro-active Behavior in Evolving Systems

Goce Trajcevski* and Peter Scheuermann**

Northwestern Univ., Dept. of EECS
{peters,goce}@eecs.northwestern.edu

Abstract. One distinct characteristics of the context-aware systems is their ability to react and adapt to the evolution of the environment, which is often a result of changes in the values of various (possibly correlated) attributes. Based on these changes, reactive systems typically take corrective actions, e.g., adjusting parameters in order to maintain the desired specifications of the system's state. Pro-active systems, on the other hand, may change the mode of interaction with the environment as well as the desired goals of the system. In this paper we describe our $(ECA)^2$ paradigm for reactive behavior with proactive impact and we present our ongoing work and vision for a system that is capable of context-aware adaptation, while ensuring the maintenance of a set of desired behavioral policies. Our main focus is on developing a formalism that provides tools for expressing normal, as well as defeasible and/or exceptional specification. However, at the same time, we insist on a sound semantics and the capability of answering hypothetical "what-if" queries. Towards this end, we introduce the high-level language $\mathcal{L}_{\mathcal{EAR}}$ that can be used to describe the dynamics of the problem domain, specify triggers under the $(ECA)^2$ paradigm, and reason about the consequences of the possible evolutions.

1 Introduction and Motivation

ER diagrams [8] have provided a foundation for a variety of tools for communicating between the domain-experts and system developers, which greatly facilitates the design process of the applications. As a typical example, the first-step in designing a database for a particular enterprize involves an ER-based description of the tables, their attributes and constraints [28]. However, recent technological advances have opened application domains in which the data items are generated by distributed and heterogeneous sources, and change very frequently, arriving in a stream-like manner [6]. These data-properties have spurred extensive research efforts in several fields. In Event-Notification Systems (ENS), and Publish-Subscribe (P-S) systems [4], typically an instance user's profile is

---

* Research supported by the Northrop Grumman Corp., contract: P.O. 8200082518.
** Research supported by the NSF grant, contract: IIS-0325144/003.

P.P. Chen and L.Y. Wong (Eds.): ACM-L 2006, LNCS 4512, pp. 30–46, 2007.
© Springer-Verlag Berlin Heidelberg 2007

matched against the current status of continuously evolving data sources, and appropriate notifications are sent to the user. Similarly, the main focus of Continuous Queries (CQ) processing [7,17] is on efficient management of user queries over time, without forcing the users to re-issue their queries. The data values may arrive as streams which the system has to process on the fly [6,19] and, furthermore, the data may be multidimensional in nature, as is the case in Location-Based Services (LBS) [21] and Moving Objects Databases (MOD) [14]. In particular, in sensor networks' settings, the data management must consider other constraints such as the limited battery-lifetime of the nodes [32].

One may observe that in the majority of the applications, there is a need for some form of a *reactive* behavior. The database community has provided many results on the topic of Active Databases (ADb), which manage triggers operating under the Event-Condition-Action (ECA) paradigm [11,20,30]. In the recent years there have been works incorporating ECA-like triggers in novel, highly-heterogeneous, distributed and dynamic data-driven application domains, e.g., the Web [13], peer-to-peer (P2P) systems and sensor networks [32]. Despite the co-existence of the large body of works in ENS, CQ, Data Streams, MOD [4,7,17,21,14], all of which have the common need of dealing with dynamically changing information, and the rich history of ADb results [11,20,30] – there is a lack of tools that would enable using the "best of all the worlds". Namely, there is no paradigm that allows the users to seamlessly tie: (1) Detection of (composite) events obtained by monitoring continuously changing data with (2) Evaluation of conditions that are continuous queries and with (3) Dynamical adjustment of the triggers themselves – all for the purpose of executing a desired policy in a constantly evolving domain of interest.

In this paper, we present our results regarding the paradigm for specifying the evolving reactive behavior that has pro-active consequences [27] and we describe our vision about the tools that would enable hypothetical reasoning and active learning about the behavior of the systems which implement such functionality. At the heart of our motivation are the following observations:

1. Whorf's hypothesis of psycholinguistics [29], which states that *"...the language that a person uses to describe his or her environment is a lens through which he or she views that environment..."*. This reflects the fact that when many domain experts are involved in the specification stage, there is a need for tools that provide as much flexibility for expressing the domain knowledge as possible.

2. During the early design stages [3], users are typically interested in *sound* but not quite *complete* descriptions, but would still like the ability to test the *behavioral correctness* of the system [9].

3. Aside from ordering among the activities in "mostly normal" executional scenarios, the users need the ability to specify exceptions, defeasibility, and reason about their impact on the system's behavior. Furthermore, these abilities are needed even when the system is deployed, preferably at minimal expense (consumption) of system's resources [15].

In the rest of this paper, in Section 2 we describe our $(ECA)^2$ (Evolving and Context-Aware Event-Condition-Action) paradigm for specifying the pro-active reactive behavior and we describe the concept of the meta-triggers which is used to orchestrate the reactive behavior in distributed settings [27]. In Section 3 we outline the syntax of the high-level language $\mathcal{L}_{\mathcal{E}\mathcal{A}\mathcal{R}}$ that can be used as a formal foundation for the desiderata listed above, and we briefly discuss its declarative semantics and its usage for hypothetical reasoning. Section 4 we presents some experimental evaluations of the benefits of the meta-trigger when managing the reactive behavior in sensor networks' settings. Section 5 positions our work with respect to the related literature and Section 6 summarizes the paper and outlines the directions of our current and future work.

2   Evolving Reactive Behavior

In order to illustrate our motivation better, we first present a scenario for which we analyze the requirements. Consider the following request:

I. Rq1: "When a moving object is continuously_moving_towards the region R for more than 5 minutes, if there are fewer than 10 fighter jets in the base B1, then send `alert_b` to the armored unit A1. Also send `alert_a` to the infantry regiment I1, when that object is closer than 3 miles to R, if all the marine corps units are further then 5 miles from R".

Rq1 needs to detect a composite event *(moving continuously towards...)*, using the individual *(location,time)* updates as simple events [26], which can be obtained, e.g., by tracking sensors [32]. **RQ1** also needs to initiate a continuous query at a remote system – the one monitoring the status of the air-base $B1$. However, **Rq1** has some subtleties: it needs the status of the air-base $B1$ for as long as the original enabling event *moving towards* is still valid. After detecting its enabling event, **Rq1** requires that the system "spans" its attention to monitoring one more event *(closer than 3 miles to R)* and, upon its detection, request an evaluation of another remote condition-query, which happens to be instantaneous. Observe that there are *bindings* between the new event to the original event – the new one needs to focus on the distance pertaining to the *particular object* that satisfied the original enabling event.

2.1   The $(ECA)^2$ Paradigm

Now we explain the main aspects of the syntax for specifying triggers under the $(ECA)^2$ paradigm, and the general form is illustrated in Figure 1. Firstly, observe that in the events, conditions and actions we allow variables to be used. For example, $E_p(EV_p)$ denotes that the event of the parent-trigger E_p has the (vector of) variable(s) EV_p in its specification; similarly, $C_{p1}(VC_p1)$ denotes the query and the variables used in the first condition of the parent trigger. We assume that the usual rules for *safety* [28] of the variables apply, in the sense that each variable that appears in a negative literal, must also appear in a positive literal, or have a ground value at the time of the invocation/evaluation of the corresponding (negated) predicate. Secondly, observe that we allow two

Fig. 1. Evolving Triggers Specification

types of children-triggers to be specified within the scope of a given (parent) trigger. As is commonly done in the programming languages, we use rectangles to visualize the nesting of the relative scope of children-triggers within the scope of the parent-trigger. As indicated in Figure 1, the user can specify an arbitrary level of nesting of descendants within the children-triggers.

Figure 2 illustrates the $(ECA)^2$ trigger that specifies the behavior expressed in **Rq1**. By using the same variable O in the events for both the *parent* and the *child* trigger, we specify that the system should begin monitoring the proximity of the particular object who has already been detected to move *continuously towards R*. Once the event describing the desired proximity criteria *"...closer than 3 miles..."* is detected, the *child* trigger begins evaluating its condition, co-existing with its *parent* trigger. The meaning of the rest of the components in the syntax of the $(ECA)^2$ triggers (c.f. Figure 1) is as follows:

1. The option **validity** in the trigger's specification allows the user to state *how long* should the trigger be considered "alive". It reflects the user's policy, and it can be either an explicit time-value, or an event which, when detected causes the particular trigger's instance to be disabled. As a special case, one is able to specify *for as long as the original enabling event is valid*, by utilizing proper expressions of the available event algebra.

2. The *Else-If* parts at each level of nesting of the triggers correspond to alternative policies, based on the value of the respective conditions, once an instance of

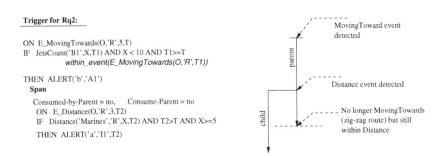

Fig. 2. Example Trigger for **Rq2**

a particular trigger is "awaken". These can be written as conditions of separate triggers with the same enabling event, but some commercial systems conforming to the ANSI standard (e.g., Oracle 9i [1]) do not allow ranking among the triggers. We do consider the option of an explicit numeric priority specification for the triggers, which can be straightforwardly extended to the conditions.

3. Each condition has two options for indicating for how long its corresponding continuous query should be evaluated. One option is to explicitly list a time-interval value, as commonly done in CQ systems (e.g., [7]). In the case of **Rq1**, the user is interested in getting updates about the state of the air-base for as long as the composite event $E\_moving\_towards$ is satisfied [26].

4. There are two types of *child-triggers*:
4.1. The first type – *child'*, enables a reaction to subsequent occurrences of other events that could potentially request monitoring of other conditions. This is the case in **Rq1**, where the user is also interested in detecting the proximity of that particular object to the region R. The value *Consumed-by-Parent = yes* indicates that the child-trigger should terminate when the parent-trigger terminates. Conversely, *Consumed-by-Parent = no*, specifies that the child-trigger should continue its execution even though the parent has ceased to exist.
4.2. The second type of a child-trigger – *child"*, is specified with the **Subsequently** option, and its intended meaning is that, after the particular parent-trigger has been enabled, and all its "options have been exhausted" (e.g., expiration of the interval of interest for the continuous 6 queries; no occurrence of the events for the *child'*-triggers), the user wants to focus on other aspects of the evolutions of the domain. This is achieved by the statement *Consume-Parent*. *Consume-Parent = yes* reflects the user's intention not to consider the parent-trigger in the future at all. In a sense, this is an equivalent to the SQL `drop` trigger rule, as no further instances of the parent-trigger are desired. *Consume-Parent = no* has the opposite effect.

2.2   Meta-triggers (and Condition vs. Event Duality)

The *meta-trigger* is a module that is in charge of several aspects and, before we go into details of explaining its roles, we present a motivational example. Recall

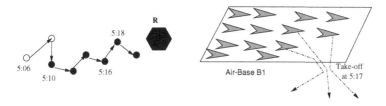

Fig. 3. Dynamics of Events and Conditions

the request **Rq1** in which the system detects the event of an object *continuously moving towards* the region R for at least 5 minutes, based on the monitoring of the object's *(location,time)* updates. Assume, for the sake of argument, that those updates are detected every two minutes.

As illustrated in Figure 3, the system began monitoring the object at 5:06, however, the *(location,time)* updates at 5:06 and 5:08, depicted with blank circles, were discarded because they were of no use for detecting the event of interest (c.f. [26]). Starting at 5:10, the system can detect the occurrence of the desired composite event at 5:16 which, in turn, "awakes" an instance of the corresponding trigger for **Rq1**. Upon checking the condition (*less than 10 airplanes in B1*), the system will find out that there are actually 12 airplanes there and, consequently, the action part *(alert)* will not be executed. Subsequently, in a very short time-span, three jets have left the air-base and, by 5:17 it has only 9 jets available. Intuitively, the trigger for the **Rq1** should fire and raise the alert. However, this may not happen until 5:18 at which time the event *moving towards* is (re)detected. In mission-critical applications, this may cause unwanted effects. Clearly, one possible solution is to periodically poll the remote database, however, this may incur a lot of unnecessary communication overhead[1] and, moreover, may still "miss" the actual time-point at which the condition became satisfied.

One of the tasks of the meta-trigger is precisely the management of this type of behavior for the continuous queries of the triggers. To specify it more formally, consider the following (simplified version of a) trigger:

TR1: ON E_1
  IF $C_{1i} \wedge C_{1c}$ **THEN** A_1

Its condition part consists of two conjuncts: C_{i1}-instantaneous[2] and C_{1c}-continuous. The meta-trigger, upon receiving the specification of the trigger from the user, identifies the instantaneous and continuous components of the condition, and does the following activities:

---

[1] Observe that the user may insist on a particular frequency of re-evaluation of the continuous query (c.f. [7]), however, we are not focusing on that issue.

[2] Observe that the evaluation of C_{1i} may be bound to various states (e.g., the one in which the event E_1 was detected; the one in which the other conjunct C_{1c} becomes true; etc.). Such bindings have already been identified as a *semantic dimension* of the active databases [11,20] and there are syntactic constructs that can specify particular states for evaluating the condition of the triggers (e.g., *referencing old/new*).

1. Translates the specifications of the original trigger into:

TR1': ON E_1 ; $(E_{C_{1c}}^{\rightarrow}$; $E_{C_{1c}}^{\leftarrow})$
     IF C_{1i} **THEN** A_1

2. It defines two new events with the Event-Base (EB):

2.1 $(E_{C_{1c}}^{\rightarrow})$ – an internal primitive event which is detected when the request for evaluating C_{1c} has been executed, which may have to be sent to a remote site – e.g., in the case of **Rq1** it is send to the air-base.

2.2 $(E_{C_{1c}}^{\leftarrow})$ – a primitive event that can be either external or internal, depending if the evaluation of C_{1c} is done locally or in the remote site. This event denotes that the condition has been evaluated to true, and the notification about it, from the possibly remote site, has been received.

3. It translates the original request for the query C_{1c} into:

3.1 A request (message, if remote site is employed) for immediate evaluation and notification if true;

3.2 A trigger that is either registered locally, or transmitted to the remote site, which essentially states:

TR1c: ON E_1; $C_{1c}^{\rightarrow}$
     IF C_{1c} valid_until($\neg E_1$)   **THEN**   A(Notify($E_{C_{1c}}^{\leftarrow}$))

4. Lastly, the meta-trigger generates the specification of another local trigger, whose description we omit – but whose purpose is to detect when the original trigger **TR1**, as specified by the user, has "expired" (e.g., the enabling composite event becomes invalidated), and:

4.1. `dissable` and/or `drop` the local triggers

4.2. Generate a notification that **Tr1c** can be `dissabled` and/or `droppedd`

What we described above exemplifies how something that was initially perceived as a pure query-like condition, becomes a "generator" of a few new events/triggers. We only explained the basic functionality of the meta-trigger as a translator for (a simplified version of) the original specifications of the user's trigger. Clearly, in reality, one may expect more sophisticated queries whose translation and generation of the equivalent new events, triggers, and messages to the remote sites will be more complicated. However, in the settings of the **Rq1**, one may observe another motivation for translating the original condition's query: the predicate `JetsCount` (c.f. Figure 2), say, for security purposes, may be a *view* and the user cannot express much at the specification time of the corresponding trigger.

At the current stage, the meta-trigger is a module that is loosely coupled with several "regular" components of a DBMS with active capabilities which, in our case happens to be Oracle 9i [1], and its details are presented in [27]. Once the user submits the specifications of the trigger (i.e., `create or alter` it), the meta-trigger will perform the necessary translations and generate the appropriate new events and triggers. The other main functions of the meta-trigger module are:

• Ensure the proper ordering for the execution of the triggers, in accordance with the given priorities (c.f. Section 3.1). We re-iterate that the ordering is not a property that is present in some commercial DBMS systems [1].

Actually, ordering of the triggers execution was our original motivation for writing a PL/SQL implementation of what became the concept of the meta-trigger: in [23], the triggers' order is dynamically changed for the purpose of minimizing the response time of updating the answers to continuous range queries, when the trajectories experience abnormalities (e.g., accidents on given road-segments).
• Ensure that, upon "ceasing" of a particular trigger, the appropriate clean-up action is taken, both in the sense of `Dissable` and `Remove`. Furthermore, the meta-trigger needs to `Enable` the respective children-triggers at the appropriate state of the evolution of the system and, when needed, `Dissable` their instances and/or `Remove` them.

3   Domain Description and Reasoning

To provide a tool for specifying the behavior of dynamic systems we propose the language $\mathcal{L}_{\mathcal{EAR}}$, which is based on action theories. Similar languages have been successfully used for specifying parameterized actions, qualification and ramification constraints, concurrent execution of actions and facts exempt from common-sense law of inertia, and reasoning about robot control programs [12,22]. We assume that there are four, possibly (countably) infinite, pairwise disjoint sets of symbols: $\mathcal{A}$, $\mathcal{F}$, $\mathcal{E}$, and $\mathcal{R}$; and a set of variables. Each symbol has an *arity* associated with it and predicates/literals from each set are defined as usual (c.f. [28]). Atoms from $\mathcal{A}$, $\mathcal{F}$, $\mathcal{E}$ and $\mathcal{R}$ are called *actions, facts, events,* and *rules* respectively. For brevity, in the sequel we exclude the variables from the presentation.

• *Effects* – We assume that an execution of a particular action in an environment in which certain facts and events hold, will modify a given fact:

$$a \textbf{ causes } f \textbf{ if } p_1, \ldots, p_n, e_1, \ldots, e_m \tag{1}$$

The intuitive meaning is that the execution of the action a in a in which each fact $p_i(i \in \{1, \ldots, n\})$ is true and each (primitive and/or composite) event $e_j(j \in \{1, \ldots, m\})$ has been detected, ensures that the fact f is true in the state resulting from the execution of the action.
• *Views* – this proposition specifies a definition of a view which, in a sense, describes the ramifications of executing an action in a given state:

$$p_1, \ldots, p_n, e_1, \ldots, e_m \textbf{ suffice\_for } q \tag{2}$$

• *Event Definitions* – We have three kinds of propositions related to events definitions:

$$a \textbf{ induces } e \textbf{ if } e_1 Y_1, \ldots, e_m, q_1, \ldots, q_n \tag{3}$$

The execution of the the action a in a state in which each of the facts q_i is true and each of the event e_j is detected, generates the event literal e, which is added to the EB if it is is positive, or deleted if the event literal e is negative. The second kind of proposition related to event definition is of the form:

$$EXPR_{events} \textbf{ induces } e_{composite} \textbf{ if } e_1, \ldots, e_m, q_1, \ldots, q_n \tag{4}$$

This propositions allow specification of the detection of a composite event in a particular state, which is defined with the $EXPR_{events}$ expression in the respective Event Algebra [2] provided by the system. The last kind of event proposition allows the user to specify the policy for consuming primitive-constituent events when a composite event is detected:

$$e_{composite} \text{ \textbf{consumes} } [e_1, \ldots, e_k] \tag{5}$$

• *Definitions of Active Rules* – Let $\alpha = [a_1, a_2, \ldots, a_n]$ denote a sequence of actions, whose effects are captured by the respective propositions of the kind (1)-(5). An active rule in the context of the $(ECA)^2$ paradigm is the proposition:

TRr: ON e_t **<priority>**   **validity(T/E)**
   if $p_1, \ldots, p_n$ **within_time/event(T1/E1)**
   [**initiate** α]

    . . .

   ELSE   **Consumed-by-Parent = <yes/no>**
    ON e_{c1}
    if $p_1^{c11}, \ldots, p_n^{c1n}$
    initiate $[\alpha_{c1}]$

     . . .

   Subsequently:
    Consume_Parent = <yes/no>
    ON e_{c2}
    if $p_1^{c21}, \ldots, p_n^{c2n}$
    initiate $[\alpha_{c2}]$ (6)

3.1   Declarative Semantics of $\mathcal{L_{EAR}}$ and Hypothetical Reasoning

Let $D_{\mathcal{EAR}}$ denote a set of ground instances of the propositions providing the domain description. We will refer to any set of facts as a *fact state*, denoted as Φ, and any set of events as an *event state*, denoted as Σ . We say that a fact f holds in a fact state Φ if $f \in \Phi$, and $\neg f$ holds in Φ if $f \notin \Phi$. Similarly, an event e holds in an event state Σ if $e \in \Sigma$, and $\neg e$ holds in Σ if $e \notin \Sigma$. Let Π denote the set of active rules which are *enabled* (i.e., their triggering event is present in the current state), and let Θ denote the set of enabled triggered rules which are available eligible for consideration (evaluation of the conditions), based on the partial ordering of their priority values. Finally, let Ω be the set of considered rules whose condition evaluated to *true* and are ready to be fired (i.e., to have the sequence of their action executed). We refer to the quintuple of the form: $S = < \Phi, \Sigma, \Pi, \Theta, \Omega >$ as a *state* of the system.

 The central concept in the declarative semantics of the $\mathcal{L_{EAR}}$ are the definitions of transition functions called causal interpretations. A *causal interpretation* is a partial function Ψ that maps a (possibly empty) sequence of actions α and a state $< \Phi, \Sigma, \Pi, \Theta, \Omega >$ into a new state. Given a domain description D, we would like to identify the causal interpretations that model the behavior of D given any initial state. We will do that through four auxiliary functions that will describe how an action, when executed in a state $< \Phi, \Sigma, \Pi, \Theta, \Omega >$, affects each component of the state. We will also need an action selection function. An *action selection function* S is a total function that takes a set of events Σ and a set of considered rules Ω, and returns the sequence of actions appearing in some rule r_i in Ω, such that action execution event $e_a^{r_i}$ of r_i is in Σ. If such a rule does not exists it returns a special *null* action μ. Each selection function S has an associated function S' that when applied to Σ and Ω, returns a singleton set with the rule $\{r_i\}$ which contains the sequence $S(\Sigma, \Omega)$ if it is not the *null* action; otherwise it returns an empty set. Action selection functions will be used to determine which actions' sequence will be selected for execution when several active rules in D are ready to be executed, based on the relative priority of the rules.

Although we do not consider any explicit "transactional awareness" which is part of the semantic dimensions of the triggers in traditional active database systems [11], we do assume (implicitly) the existence of *rules' processing point* in every legal sequence of actions. Observe that if S selects a sequence of actions which does not have a processing point at the end of the list, no new rules will be allowed to fire at the end of executing the selected sequence (i.e. the rules in Ω will have to wait until a new processing point is encountered). With minor modifications to the definition of models we could assume that rules by default are processed each time we get into an empty sequence of actions (so that rules will be processed at least at the end of a given transaction) in addition to the explicit processing points. Furthermore, we can put a restriction on a syntax which will require that every sequence of actions must end with a $\uparrow$ symbol – denoting special processing points.

The hypothetical reasoning about the behavior of the system is based on the simple *query language* which consists of the statements of the following form:

$$l \text{ after } \alpha \text{ at } < \Phi, \Sigma > \qquad (7)$$

where l is either a *fact* or an *event literal*, α a sequence of actions, and $< \Phi, \Sigma >$ a state of facts and events. For a query q, we will denote by $\neg q$ the query $\neg l \text{ after } \alpha \text{ at } \text{¡} \Phi, \Sigma >$. We use $Cn(D_{\mathcal{EAR}})$ to denote the set of all the facts entailed by the $D_{\mathcal{EAR}}$ description and a given state.

By asking queries like (7), one can actually reason about the consequences of executing a sequence of actions in a given state of the system, and the formal tool for reasoning is based on:

Definition 1. We say that a query q of the form (7) is true in a model Ψ of a domain description $D_{\mathcal{EAR}}$ iff l holds in the fact state of the state $\Psi(\alpha, < \Phi, \Sigma, \emptyset, \emptyset >)$. A description $D_{\mathcal{EAR}}$ *entails* a query q (written as $D_{\mathcal{EAR}} \models q$) iff q is true in all models of $D_{\mathcal{L_{EAR}}}$.

As an example, our tool enables the user in **Rq1**-like settings to pose queries such as: *"Will I have sufficient defense resources if 20 objects are continuously moving towards R for more than 5 minutes within half an hour, approaching from North-West, and three of them are within 3 miles"*, and get the answer based on his description of the environment. This could be expressed as:

(Count(Airplanes(F15,B1)) > 5 AND Count(Infantry(A1,Marksman) > 17))
after
{ [Moving_towards(X1, 'R', 'NW',5, T1) AND Moving_towards(X2, 'R', 'NW',5, T2)
AND ... Moving_towards(X20, 'R', 'NW',5, T20)] AND (T1< T2 ... < T20) AND
(T20 - T1 < 30) AND [(Distance(Xi, 'R') < 3) AND (Distance(Xj, 'R') < 3) AND
(Distance(Xk, 'R') < 3)] }

3.2 Defeasibility and Exceptions

We are interested in *expected* and *unexpected* categories of failures [5]. Expected exceptions correspond to the executions which are not desirable but may happen (hopefully, very rarely). Since the designers may not be aware of all the possible executional scenarios, unexpected exceptions may need to be handled. Hence, we need to allow for the activities (i.e., the consequences of their executions) and some ramifications to be *defeasible*, in a sense that their effects do not apply in a particular state s because that state is considered to be exceptional. To handle this exceptional behavior, we introduce the following propositions:

$p_1, \ldots, p_n, e_1, \ldots, e_m$ **exceptionally_suffice_for** f
a **exceptionally_causes** f **if** $p_1, \ldots, p_n, e_1, \ldots, e_m$
a **exceptionally_induces** e **if** $p_1, \ldots, p_n, e_1, \ldots, e_m$
a **exceptionally_determines** f **if** $p_1, \ldots, p_n, e_1, \ldots, e_m$

In the exceptional situations, the effects of the activities are determined by the exceptional propositions and the effects of the defeasible propositions are ignored.

4 Experimental Observations

Based on the benefits of the ordering of triggers execution, demonstrated in [10,23], we conducted simulation-based experiments to demonstrate the benefits of the meta-triggers when processing a reactive behavior in distributed environments. To do so, we considered the following request:

Rq2: *"When more than 50% of the sensors in the region R1 report temperature readings above 60C (continuously) for more than 5 minutes, if more than 70% of the sensors in the region R2 continuously report the readings of the CO concentration that are above 20% for 3 minutes, then double the sampling frequency of the sensors in R2 and alert the security vehicles."*

If a meta-trigger is not employed, the system is bound to evaluate the continuous queries via constant polling. To evaluate the benefits of the push-based notification, we used our sensor network simulator SNSim[3] described in [25]. We used it to observe the number of messages exchanged when processing the request **Rq2**. For that purpose, we randomly generated pairs of rectangular regions $R1$ and $R2$ (a user can do it via a graphical interface), in each of which

---

[3] Source code available at *http://www.eece.northwestern.edu/peters/research/Sensors*

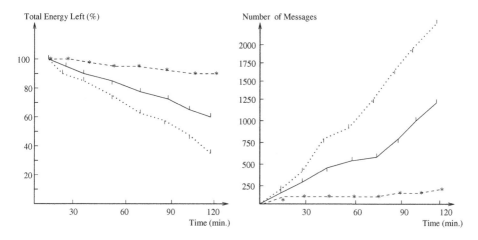

Fig. 4. Effects of the meta-trigger on the efficiency

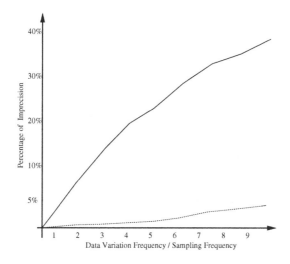

Fig. 5. (Im)Precision Effects of the Meta-Triggers

the node closest to the centroid of the respective rectangle was selected as a
cluster-leader. We varied the following parameters: – the frequency of changes
in the values read by each sensor; – the magnitude of changes in the values read
by each sensor; – the polling frequency that the sink uses to query the readings
in $R2$ when the event is detected in $R1$. The experiments were performed on
an Intel Pentium IV CPU 3.8GHz processor, with 1G of DDR2 memory. Our
network consisted of 400 nodes whose parameters emulated the MICA-Motes,
and we executed 50 different $(R1, R2)$ pairs, running the simulation for 2 hours
for different combinations of the parameters above. Our observations are depicted in Figure 4 where the left portion represents the energy lost in the nodes,

and the right portion depicts the number of messages exchanged. The solid line represents the average of all observations operating without the meta-trigger installation in the cluster leaders of $R1$ and $R2$, $R2$ being constantly polled. The dotted line depicts our worst-case observations. However, as indicated with the dashed line, when meta-triggers are used, there is an order of magnitude of savings in the messages exchanged and, conversely, much larger energy savings.

The other set of the experiments was concerned with the (im)precision issues. Namely, if periodic polling is used, the system may fail to detect changes in the values of the condition (continuous query) if those changes happen fast and in-between two consecutive evaluations. The system may fail to detect the condition even if the meta-trigger is used, but this is only due to a communication delay and in the extreme case when it occurs at the expiration of the enabling event (which is also possibile when polling is used). Our results are depicted in Figure 5. The x-axis denotes the ratio of the frequencies of changes in the values that are monitored by individual sensors and their sampling frequency; y-axis is the percentage of misses in detecting that the condition has become true. As can be seen, the imprecission ("misses") is much higher without a meta-trigger.

5   Related Works

Systems with (re)active behavior can be found in different research fields: active databases [31], monitoring and streaming sensor data [18], publish/subscribe (e.g., Siena [4]) and event-based systems (e.g., A-mediAS [16]). However, the topic of adaptivity in the observations and triggers has rarely been addressed.

Composite events for active database systems have been studied for a long time, also with special focus on composite events (for overview see [33,11]). Our notion of evolving triggers and conditions can be seen as the dynamic extension of the concept of coupling modes: there, the relationship between condition and trigger is described in relation to the transactions the rule is executed in. Active database systems can rely on the transactional context for the composition of events, which is not the case in all of the application fields for which our approach has been developed. In addition, the issues of event consumption have been discussed at length for composite events in active database systems. In typical ADb the mode is often set implicitly for a given ECA rule; extended ECA (EECA) rules made the semantic dimensions explicit. The introduction of interval-based semantics (e.g., in [2]) addressed the issue of validity of triggers for a certain interval. This approach is related to ours (as the trigger execution is depending on a change) but does not aim for the same goals – the evolution of the triggers and, consequently, the techniques are not directly applicable to our work.

Changing requests over streaming data are considered in [6] and the main focus is on adding and updating new/changed requests online. Few ENS support composite events; most services are message centered and do not rely on an event model. For example, Siena [4] supports only simple composite events without conditions or parameters. Event requests in the continuous query project [7,17] support sophisticated composite events. They consist of triggering events and conditions and allow specification of start/end-times for the trigger evaluation. These structures could be extended in a way that they consider changes (as in our case). Meta-ENS [16] adapt algorithms to changing distributions and translate semantics between services, respectively. Currently, these projects do not consider the continuous interdependency between triggering events and conditions. As shown in the examples, our approach is particularly useful when considering mobile targets. In recent years, there has been an increasing interest in systems that consider mobile data sources and location-based systems [21]. Most systems are still in a fledgling stage, where the influence of location is rather direct and complex dependencies on combinations of location, direction, other contexts several mobile peers and static items are only just evolving.

6   Concluding Remarks and Future Work

We presented our $(ECA)^2$ paradigm for specifying triggers that are aware of the dynamic correlation between the events and conditions and, in a sense, can "react in a proactive manner" – by modifying themselves. We also introduced the syntactic constructs for a high-level language $\mathcal{L_{EAR}}$ that can be used for specifying the triggers under this paradigm in dynamic distributed environments. Based on the declarative semantics of $\mathcal{L_{EAR}}$, we also presented a tool that can be used for hypothetical reasoning regarding the behavior of dynamic system in incomplete-specification settings. In addition, we also discussed the option of providing constructs for specifying the propositions that describe the consequences of exceptional behavior, in addition to the "normal" (expected) behavior.

Currently, we are developing the translator for a subset of the functionality of the active rules specified in the $\mathcal{L_{EAR}}$ style, that we would like to incorporate on top of an off-the-shelf commercially available DBMS (Oracle). Similarly to [11] we are also building an Event-Base, a separate module for managing the detection of composite events and the consumption of the primitive ones. We are also working on further incorporating the $(ECA)^2$ in heterogeneous/multidatabase settings, and we are extending our SNSim simulator for sensor networks (c.f. [24]) with the reality-awareness.

Our vision is to implement a system that can provide the functionalities described in the previous section, and its main components are presented in Figure 6.

Firstly, we would like to be able to automatically translate between visual (graph-based) specifications of the sequencing of the activities, and their corresponding textual counter-parts. Clearly, a correctness of the mutual translation will need to be guaranteed, however, this is not the only interaction between the two components. Namely, based on the domain-description propositions, when using the graphical portion, the user will need to be automatically prompted to provide values for certain parameters (e.g., the *actor* for a given role). Our earlier work [22] has provided somewhat similar functionalities in the context of workflow systems, however, it did not incorporate any management of a reactive behavior.

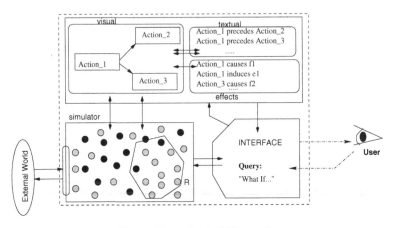

Fig. 6. Hypothetical Reasoning

An important component of our envisioned system is the *reality-aware simulator*. This module serves the purpose of answering the users queries while minimizing the access to the actual reality modelled by the system. In particular, Figure 6 illustrates a scenario in which the user is interested in the energy-map in the region R of a given sensor network. Actually querying the network itself about the energy status of the nodes will impose a computational overhead that will even further drain the batteries of the individual nodes. Hence, if the representation in the simulator can provide satisfactory Quality-of-Data (QoD) guarantees, the user's request can be answered without accessing the real environment at all. Towards this, it is a paramount that the simulator updates its knowledge about the parameters of the environment by properly reacting to certain events and, moreover, steering its reactive behavior in the future.

What this would accomplish is that even at the run-time, the user can still ask hypothetical queries of the form: *"Will the sensors be able to track 3 objects inside the region R for 2 hours?"*.

References

1. Oracle 9i. www.oracle.com/technology/products/oracle9i
2. Adaikkalavan, R., Chakravarthy, S.: Formalization and detection of events using interval-based semantics. In: COMAD (2005)
3. Bergamaschi, S., Sartori, C.: On taxonomic reasoning in conceptual design. ACM Transactions on Database Systems 3(17) (1992)
4. Carzaniga, A., Rosenblum, D.S., Wolf, A.L.: Design and evaluation of a wide-area event notification service. ACM-TOCS 19(3) (2001)
5. Casati, F., Pozzi, G.: Modeling exceptional behavior in commercial workflow management systems. In: 4th Intl. Conf. on Cooperative Information Systems (CoopIS) (1999)
6. Chandrasekaran, S., Franklin, M.: Streaming queries over streaming data. In: VLDB Conference (2002)

7. Chen, J.J., DeWitt, D.J., Tian, F., Wang, Y.: Niagaracq: A scalable continuous query system for internet databases. In: ACM SIGMOD Conference (2000)
8. Chen, P.P.: The entity-relationship model: Towards a unified view of data. ACM-TODS 1(1) (1976)
9. Davulcu, H., Kifer, M., Pokorny, R.L., Ramakrishnan, C., Dawson, S.: Modeling and analysis of interactions in virtual enterprises. In: Research Issues on Data Engineering (RIDE) (1999)
10. Ding, H., Trajcevski, G., Scheuermann, P.: Omcat: Optimal maintenance of continuous queries answers for trajectories. In: ACM SIGMOD (2006) (demonstration paper)
11. Fraternali, P., Tanca, L.: A structured approach for the definition of the semantics of active databases. Transactions on Database Systems, 20(4) (1995)
12. Giunchiglia, E., Lifschitz, V.: An action language based on causal logic. In: AAAI 1998 (1997)
13. Papamarkos, G., Poulovassilis, A., Wood, P.T.: Event-condition-action rule languages for the semantic web. In: Workshop on Semantic Web and Databases (at VLDB) (2003)
14. Güting, R.H., Schneider, M.: Moving Objects Databases. Morgan Kaufmann, San Francisco (2005)
15. Hull, R., Llirbat, F., Simon, E., Su, J., Dong, G., Kumar, B., Zhou, G.: Declarative workflows that support easy modifications and dynamic browsing. In: Intl. Joint Conference on Work Activities Coordination and Collaboration (WACC) (1999)
16. Jung, D., Hinze, A.: A meta-service for event notification. In: CoopIS/DOA/ODBASE (1) (2004)
17. Liu, L., Pu, C., Tang, W.: Continual queries for internet scale event-driven information delivery. IEEE-TKDE 11(4) (1999)
18. Madden, S., Shah, M.A., Hellerstein, J.M., Raman, V.: Continuously adaptive continuous queries over streams. In: ACM SIGMOD Conference (2002)
19. Olston, C., Jiang, J., Widom, J.: Adaptive filters for continuous queries over distributed data streams. In: ACM SIGMOD (2003)
20. Paton, N. (ed.): Active Rules in Database Systems. Springer, Heidelberg (1999)
21. Schiller, J., Voisard, A.: Location-based Services. Morgan Kaufmann, San Francisco (2004)
22. Trajcevski, G., Baral, C., Lobo, J.: Formalizing and reasoning about the requirements specifications of workflow systems. IJCIS 10(4) (2001)
23. Trajcevski, G., Ding, H., Scheuermann, P.: Context-aware optimization of continuous range queries for trajectories. In: MobiDE Workshop (2005)
24. Trajcevski, G., Ghica, O., Scheuermann, P.: Car: Controlled adjustement of routes and sensor networks lifetime. In: MDM (2006)
25. Trajcevski, G., Ghica, O., Scheuermann, P.: Car: Controlled adjustment of routes and sensor networks lifetime. In: MDM (2006)
26. Trajcevski, G., Scheuermann, P., Brönnimann, H., Voisard, A.: Dynamic topological predicates and notifications in moving objects databases. In: Mobile Data Management (MDM) (2005)
27. Trajcevski, G., Scheuermann, P., Ghica, O., Hinze, A., Voisard, A.: Evolving triggers for dynamic environments. In: Ioannidis, Y., Scholl, M.H., Schmidt, J.W., Matthes, F., Hatzopoulos, M., Boehm, K., Kemper, A., Grust, T., Boehm, C. (eds.) EDBT 2006. LNCS, vol. 3896, Springer, Heidelberg (2006)
28. Ullman, J.D.: Principles of Database and Knwoledge – Base Systems. Computer Science Press (1989)

29. Whorf, B.: Language, thought and reality. MIT Press, Cambridge (1956)
30. Widom, J.: The starburst active database rule system. IEEE Transactions on Data and Knowledge Engineering 8(4) (1996)
31. Widom, J., Ceri, S.: Active Database Systems: Triggers and Rules for Advanced Database Processing. Morgan Kaufmann, San Francisco (1996)
32. Zhao, F., Guibas, L.: Wireless Sensor Networks: an Information Processing Approach. Morgan Kauffman, San Francisco (2004)
33. Zimmer, D., Unland, R.: On the semantics of complex events in active database management systems. In: ICDE (1999)

A Common Core for Active Conceptual Modeling for Learning from Surprises

Stephen W. Liddle[1] and David W. Embley[2]

[1] Information Systems Department,
[2] Department of Computer Science,
Brigham Young Univeristy, Provo, Utah 84602, U.S.A.
liddle@byu.edu, embley@cs.byu.edu

Abstract. The new field of active conceptual modeling for learning from surprises (ACM-L) may be helpful in preserving life, protecting property, and improving quality of life. The conceptual modeling community has developed sound theory and practices for conceptual modeling that, if properly applied, could help analysts model and predict more accurately. In particular, we need to associate more semantics with links, and we need fully reified high-level objects and relationships that have a clear, formal underlying semantics that follows a natural, ontological approach. We also need to capture more dynamic aspects in our conceptual models to more accurately model complex, dynamic systems. These concepts already exist, and the theory is well developed; what remains is to link them with the ideas needed to predict system evolution, thus enabling risk assessment and response planning. No single researcher or research group will be able to achieve this ambitious vision alone. As a starting point, we recommend that the nascent ACM-L community agree on a common core model that supports all aspects—static and dynamic—needed for active conceptual modeling in support of learning from surprises. A common core will more likely gain the traction needed to sustain the extended ACM-L research effort that will yield the advertised benefits of learning from surprises.

1 Introduction

We live in an increasingly-connected and data-rich internet age where information is one of the global economy's most valuable and often-traded assets. Because of data warehouses and a variety of other information technologies (IT), companies like Wal-Mart are able to track transactions with extraordinary detail, and perform sophisticated capacity planning based on such variables as local weather forecasts, holidays, and population trends. Similarly, airlines perform IT-enabled "yield management" that successfully optimizes their use of available passenger seats and cargo capacity. Data is everywhere, and organizations of all types leverage that data to monitor the world's information state and react appropriately.

As the world "flattens" in response to technology and information accessibility [10], individuals and organizations around the world are empowered to a

P.P. Chen and L.Y. Wong (Eds.): ACM-L 2006, LNCS 4512, pp. 47–56, 2007.
© Springer-Verlag Berlin Heidelberg 2007

degree hitherto unknown. Communication and information sharing across state boundaries in the internet and web era is now relatively easy when compared with the world even twenty years ago.

However, as we empower individuals and organizations, we enable not only new opportunities, but also new threats. Our modern world is characterized by significant political unrest and sometimes open warfare (e.g., conflicts around the globe in such diverse places as Cyprus, Darfur, Iraq, Israel, Kashmir, Kosovo, and North Korea). Often this leads to surprises—such as the 11 September 2001 attacks on the United States and the 11 March 2004 Madrid train bombings—that have a major impact on local and global communities. Societies spend enormous amounts of resources on intelligence and national defense efforts in order to mitigate the threats that lead to sometimes surprising events.

Other surprises result from natural causes, such as earthquakes, volcanoes, and violent weather. Scientists study nature for a variety of reasons, but clearly some scientists are motivated by the desire to improve the condition of humanity and prevent loss of lives and property by providing early warnings of impending risks. For example, we monitor seismological activity around known areas of potentially dangerous earthquake and volcanic activity. We monitor weather patterns from the ground, air, ocean, and space in order to predict potentially dangerous storms.

It is interesting to note that "natural" surprises share many of the same characteristics as human-caused events, and so we can treat them in a similar way. We react to threats of all kinds by monitoring conditions as best we can, and putting in place policies and defenses designed to counter the threats. Unfortunately, monitoring is neither perfectly accurate nor fully preventative. Surprises have a major impact on society, and they will continue to occur. But if we can predict future events with a higher degree of accuracy, there is considerable potential benefit to society.

We agree with two premises set forth by Chen and Wong [4] regarding the goal of predicting future events that have surprised us in the past:

1. documenting and analyzing surprises using scientific theories, information technology, and conceptual modeling of data/knowledge/information can help, and
2. state-of-the-art techniques focus on static relationships and pre-defined entities of interest, rather than dynamic and time-varying relationships that would lead to better learning and predictive power.

If the goal is to be able to predict future events by learning from past experience and engaging in active, ongoing monitoring followed by scenario and risk analysis, then it is clear that traditional data modeling techniques alone are inadequate to the task. We also need to capture dynamic aspects in our conceptual models.

In our view, predictive power comes from understanding three vital elements:

1. a system's current state,
2. its likely or possible future states, and
3. its recent evolution.

The way we capture information about a system's actual or possible states is by conceptual modeling of objects, their relationships with other objects, their individual behaviors, and their interactions (collective behaviors). Questions of recent history and evolution require a temporal dimension that has been studied extensively [11], but is not common in practice.

The entity-relationship approach [2] and its principal extensions are excellent for data modeling, but do not generally provide means to model behavior [18]. However, conceptual modeling researchers do indeed recognize the benefits of moving beyond static relationships and integrity constraints to encompass dynamic, behavioral aspects of systems as well [3]. We espouse an object-oriented approach to modeling that encompasses this expanded data semantics [6]. We also advocate the use of executable conceptual models—that is, models that have fully specified formal semantics that can execute directly, allowing, but not requiring a further programming step [17].

In the remainder of this paper we examine our thesis that the conceptual modeling community has developed sound theory and practices for conceptual modeling that, if properly applied, could help analysts model and predict more accurately. Specifically, we need to associate more semantics with modeled links, and we need fully reified high-level objects and relationships that have a clear, formal underlying semantics that follows a natural, ontological approach. We argue that the research community needs an agreed-upon core set of conceptual modeling constructs for capturing dynamic aspects of systems (in addition to traditional static aspects). These concepts already exist, and the theory is well developed; what remains is to connect them with the ideas needed to predict system evolution, thus enabling risk assessment and response planning.

In Section 2 we argue for a particular set of desirable features for an active conceptual modeling foundation. Then in Section 3 we give an overview of the structure of one possible common core set of features. We offer this set as a starting point for discussion, not a final proposal. We conclude in Section 4.

2   Formal Foundation

Before we can construct a common active conceptual modeling core, we first need to establish guiding principles surrounding the framework for the core's formal definition. We set forth our recommendations in Sections 2.1 through 2.6.

2.1   Metamodeling, Formal Representation, and Mathematical Model Theory

Our experience (e.g. [15]) is that even academic researchers publishing textbooks and peer-reviewed papers tend to give less than formal definitions for most of the concepts they describe. The strong tendency is to give a natural-language description of a particular concept together with examples of the concept's use. This tendency makes sense, because most readers (including fellow academics) are mostly interested in understanding the core idea, not its formal semantics.

Unfortunately, this shortcut habit leaves ample opportunity for miscommunication and lack of real, in-depth understanding. The committee developing the Unified Modeling Language (UML) recognized this problem as well, and they attempted to resolve it by use of the Meta Object Facility (MOF) and the Object Constraint Language (OCL) to give a parallel, formal description of the UML concepts described originally in natural language. Yet as the work of Gogolla and others over the years shows (e.g. [1]), the precise semantics of UML has not been fully clear, even in its latest version. However, at least a clear mechanism (MOF and OCL) is in place to understand the nature of the problems and address them.

Today, metamodeling is widely accepted as a superior way to describe models. The main idea is to describe the model either in terms of itself or a specialized metamodeling language, and then to provide a formal definition of the metamodeling language (or model subset used for metamodeling), thus giving a formal definition of the model. Our own work follows this approach, providing a description of Object-oriented Systems Modeling (OSM) in terms of a limited subset of itself, and then giving a formal definition of that subset using mathematical model theory [6,7].

The utility of this approach manifested itself in our study of cardinality constraints in semantic data models [15], and in our work on model-equivalent programming languages [14,17]. In the first case, we had to go through textual descriptions and do our best to interpret the authors' intentions. Once we had a formal definition for each cardinality constraint, we were able to construct a mathematical lattice of all the constraints and compare the power of the various semantic data models. This yielded observations about opportunities for further development of even more powerful cardinality constraints. In the second case, we applied the technique to our own model and discovered elements of the earlier OSA model [6] that could be eliminated because they provided no additional expressive power. For example, the concepts of dominant and independent high level object classes turned out to be formally the same, even though the natural-language narrative was different for these distinct but related constructs. Thus, our refined version of OSM replaced the two with a single construct [14]. Metamodeling and formal definitions can help model designers do a better job of specifying correct and compact definitions, which in turn will help modelers by reducing the cognitive load they carry in trying to learn, understand, and use the many constructs that come with a modern model, such as UML.

Metamodeling can be helpful in designing, learning, and understanding conceptual models, but metamodeling alone is inadequate without a rock-solid formal definition. We believe that the right way to formalize a conceptual model is through mathematical model theory [8]. The main ideas in mathematical model theory are that for some language L, a model of L is a structure with a domain of discourse, a set of constants, a set of functions, and a set of n-ary relations for the predicates of L. A theory T is a set of sentences in L, and a model of T is a model of L in which all sentences of T are true. The main idea in our formalization of a conceptual model is that we first map the constructs in our conceptual model to a set of logic formulas (this is a theory, say, T), and then we consider the semantics of the conceptual model to be the set of all mathematical models for T.

2.2   Tunable Formalism

The benefit of reusing traditional model theory (as opposed to a new invention like OCL) is that it is formally understood by mathematicians and has been well studied through the years—it is the right tool for the job. However, the vast majority of the conceptual modeling community will not be interested in raw formalism, but rather will only want to deal with the conceptual model itself. The concept of tunable formalism [5] allows us to bootstrap our use of formalism and metamodeling so that we need not deal in the most rigorous mathematical detail. Since our conceptual model is formally defined, we can simply stay in the conceptual model itself as we perform our modeling tasks. We only rarely need to deal with the formal layer.

In a related note, another aspect of tunable formalism could be useful in conceptual modeling for learning, namely that OSM allows varying levels of detail and completion in a model instance. For example, a trigger might initially be written in natural language during analysis, but later would be refined to a more precise programming-type language. The particular language used to describe the trigger does not change the underlying mathematical interpretation of the model instance. If the trigger is formal, it can be executed, and if not, then it could still be simulated if a human were to guide the simulation system and indicate when the trigger is true.

2.3   Executable Models

Formal models can be executed, either by direct simulation or by generation of programming language code from formally-described model instances. One example of such a system is our OSM/Harmony prototype [14], but there are a number of other examples now as well. CARE Technologies of Spain has created a model compiler called OlivaNova Model Execution for OO-Method, which is sufficiently similar to UML that it can be viewed as a UML variant. The OlivaNova tool can generate Java, C#, or Visual Basic client/server code for three-tier solutions that correspond to OO-Method model instances. Another example is Executable UML, which is essentially a UML profile (a subset of UML) that deletes imprecise, non-executable portions of UML and supplies an action language sufficient for the purposes of real-time and embedded systems. Mellor states that Executable UML has been used on more than "1400 real-time and technical projects."

A possible benefit of executable models in a learning situation is the ease with which one could generate system simulations to test various possible future scenarios. In the past we have argued for executable conceptual models on the strength of their software engineering benefits. In this case, the expected benefit is the ability to create and test simulations quickly.

2.4   Closed vs. Open World Assumption

An interesting problem with current approaches to formal models (and databases) is that they generally rely on the closed world assumption (CWA), meaning that anything not known to be true is treated as false. CWA is fine for post hoc analysis of past surprises, but it is not obvious whether CWA is well suited for what-if analysis of an evolving system, where the goal is a priori analysis of potential threats. This question deserves further study. Alternatives include either moving to an open world assumption, or modeling unknowns explicitly in the conceptual model.

2.5   Ontological Approach

Another driving principle we advocate for conceptual modeling formalisms is the notion that terms and definitions should take a natural, ontological form. Just as it makes little sense to reinvent the wheel of mathematical model theory, so also it is inefficient for conceptual modelers if the model designers use terms in an unnatural way. For example, consider the heavily overloaded term "object." Depending on the context, it might mean (1) "a person, place, or thing, real or abstract," or (2) "a model of a real-world object," or (3) "a machine representation of an object in sense #2." An ontological approach to defining our conceptual modeling terms can avoid many of the ambiguities that arise when moving from one software development activity to another, or when working with a team of analysts [16]. An ontological perspective would adopt sense #1 for the definition of object, even though programming languages almost uniformly adopt a form of sense #3.

2.6   Fully Reified High Level Constructs

An information network to support learning from surprises will likely need to be super-massive. For network theory to help (which, as we have argued elsewhere, is useful [19]) we first need a high-level data representation construct that is fully reified, meaning that high-level and low-level constructs are fully interchangeable. OSM provides just such a notion with its high-level object set and high-level relationship set constructs.

When we mark an object set as being "high level," we allow it to contain other object sets and relationship sets. We do add the constraint that if a high-level object set contains a relationship set, it must also include all its connected object sets. This ensures that the containment boundary is clean (i.e., a contained relationship does not depend on objects outside the high-level boundary). A high-level object set is otherwise identical to an ordinary object set. It has a first-class identity that does not depend on some aggregate formula involving other objects and relationships. Members of the high-level object set can be created and destroyed in exactly the same way as other object sets, and so forth. Since we have a clean boundary and a high-level object set is fully reified, we can collapse and expand high-level components at will. If we wish to consider the details within the high-level component, we expand it. If not, we collapse it, and other than having a shaded box, it appears identical to any other object set. We can cleanly ignore the contained details because the high-level object set is a first-class object set in every way. This provides a powerful abstraction mechanism that helps us deal with considerable complexity in a straightforward manner. The case for high-level relationship sets is similar.

The ideas we have summarized in Sections 2.1 through 2.6 are not ours alone, but they are the product of conceptual modeling reseachers over the past several decades. In summary, the conceptual modeling community has built a large body of theory around mathematically sound principles that we should leverage for conceptual modeling of learning from surprises.

3   A Common ACM-L Core

The active conceptual modeling for learning (ACM-L) community is just getting started, and the ACM-L objectives are ambitious. It will take a robust, sustained effort to accomplish those objectives. Thus we suggest that the best chance for success starts with establishing a common core set of constructs that the ACM-L community agrees will be the point of departure for future research.

One of the dangers of standards is that they are necessarily the product of committee work, and there is a strong tendency toward scope creep in such efforts. UML is an example of a relatively simple idea that has turned into a truly massive specification. (The UML 2.0 Superstructure document is 710 pages of diagrams and text.) UML is the object modeling standard, but its full details are incomprehensible to the ordinary practitioner. So we add one more feature to the list of Section 2: minimalism. The ACM-L core should be nonredundant.

In accordance with these principles, the following is a short list of concepts needed to fully represent the structural aspects of a system:

- Object
- Relationship
- Object set (class)
- High-level object set
- Relationship set
- High-level relationship set
- Generalization/specialization
- Constraints
  - Cardinality
  - Generalization/specialization
  - General

The main idea is that there are objects and relationships among objects, and they are organized into sets that can be related in various ways. The following is a similarly short list of concepts needed to fully represent the dynamic aspects of a system:

- States
- High level states
- State conjunctions
- Transitions
  - Trigger
  - Action
- High-level transitions
- Exceptions
- Threads
- Constraints
- Interactions
  - Origin
  - Destination
  - Parameter list
- High-level interactions

The main idea for representing the behavior of an object is that threads can exist in various states for an active object, and transitions between states occur according to well-defined transition rules. Objects and threads communicate and synchronize through interactions.

Like these lists, the OSM metamodel is also concise, fitting comfortably on three normal pages [7,14]. Metamodel diagrams do not tell the whole story, but conceptual modeling researchers can understand this set of constructs much more readily than the expansive UML 2.0 standard.

We offer these lists as a starting point for discussion of a common ACM-L core.[1] We recommend an object-oriented approach, but we recognize that this is a point of discussion that needs to be debated within the ACM-L community. There are alternatives, such as using Petri nets [21] or Harel-style statecharts [13] instead of our state nets. An extended ER model (e.g. see [9]) or Object Role Modeling [12] could be adopted in place of our object-relationship model. What is most important is that the ACM-L community agree to a common core.

Of course we are comfortable with OSM and would be happy to see it widely adopted, but that is not our goal here. The object modeling community has moved on, and UML is the standard. Unlike OSM, UML enjoys broad tool support and has significant "mind share" among practitioners and researchers alike. So why not simply adopt UML? Unfortunately, UML's sheer size and complexity make it impractical as a common ACM-L core. For ACM-L research we need a concise set of fully precise, formal definitions. And yet UML is the standard, with all the advantages that position confers.

A potential resolution for this dilemma is patterned by the work on Executable UML [20], where a UML profile (a restricted set of UML constructs) designates the subset of UML that will be used. We should examine whether such an approach could work to provide the features of the ACM-L common core model within a UML context. The overall solution framework we envision

---

[1] Because of space constraints we do not explain these constructs in detail here. See [7] for full details.

includes a database and a data warehouse (or perhaps several) with a conceptual modeling front-end to help manage the complexity of the underlying information. We believe that solutions to this complex learning and prediction problem should go to great lengths to avoid reinventing the theories and techniques that have already been developed and are well-established, albeit in separate communities (e.g., conceptual modeling, database theory, artificial intelligence, and social networking, to name a few).

4   Conclusion

We have described the need for active conceptual modeling in support of learning from surprises. We have also explained the elements of a theoretically sound formalism for ACM-L, and have offered a beginning point for a debate on the constructs that should form a common ACM-L core set of modeling constructs. Further we suggest that it may be appropriate to define this core in terms of a UML profile, thus adhering to the industry standard while achieving the precision and formalism we require.

There is good reason to be optimistic that we can contribute to the field of predicting future surprises based on learning from the past and properly modeling the world's global state. We already have a great many tools to help, from social networking to conceptual models and database management systems, data warehouses, and OLAP. The task may require us to revisit former ideas in a new context (e.g., spatio-temporal modeling and active databases), and it will certainly be a multi-disciplinary, collaborative effort. It is our position that we have the building blocks, but that they need to be put together in a suitable way.

References

1. Bauerdick, H., Gogolla, M., Gutsche, F.: Detecting OCL Traps in the UML 2.0 Superstructure: An Experience Report. In: Baar, T., Strohmeier, A., Moreira, A., Mellor, S.J. (eds.) UML 2004. LNCS, vol. 3273, pp. 188–196. Springer, Heidelberg (2004)
2. Chen, P.P.: The Entity-Relationship Model: Toward a Unified View of Data. ACM Transactions on Database Systems 1(1), 9–36 (1976)
3. Chen, P.P., Thalheim, B., Wong, L.Y.: Future Directions of Conceptual Modeling. In: Chen, P.P., Akoka, J., Kangassalu, H., Thalheim, B. (eds.) Conceptual Modeling. LNCS, vol. 1565, pp. 287–302. Springer, Heidelberg (1999)
4. Chen, P.P., Wong, L.Y.: A Proposed Preliminary Framework for Conceptual Modeling of Learning from Surprises. In: Arabnia, H.R., Joshua, R. (eds.) ICAI 2005. Proceedings of the 2005 International Conference on Artificial Intelligence, vol. 2, pp. 905–910. CSREA Press (June 2005)
5. Clyde, S.W., Embley, D.W., Woodfield, S.N.: Tunable Formalism In Object-Oriented Systems Analysis: Meeting The Needs Of Both Theoreticians And Practitioners. In: Clyde, S.W., Embley, D.W., Woodfield, S.N. (eds.) Proceedings of the 1992 Object-Oriented Programming Systems, Languages and Applications, pp. 452–465 (October 1992)

6. Embley, D.W., Kurtz, B.D., Woodfield, S.N.: Object-oriented Systems Analysis: A Model-Driven Approach. Prentice Hall, Englewood Cliffs, New Jersey (1992)
7. Embley, D.W.: Object Database Development: Concepts and Principles. Addison-Wesley, Reading, Massachusetts (1998)
8. Enderton, H.B.: A Mathematical Introduction to Logic. Academic Press, New York (1972)
9. Elmasri, R., Navathe, S.B.: Fundamentals of Database Design, 5th edn. Addison-Wesley, Reading (2007)
10. Friedman, T.L.: The World Is Flat: A Brief History of the Twenty-First Century. Farrar, Straus and Giroux, New York (2005)
11. Gregersen, H., Jensen, C.S.: Temporal Entity-Relationship Models—A Survey. IEEE Transactions on Knowledge and Data Engineering 11(3), 464–497 (1999)
12. Halpin, T.: Conceptual Schema and Relational Database Design. Prentice Hall, Englewood Cliffs (1995)
13. Harel, D.: Statecharts: A visual formalism for complex systems. Science of Computer Programming 8(3), 231–274 (1987)
14. Liddle, S.W.: Object-Oriented Systems Implementation: A Model-Equivalent Approach. PhD Dissertation, Department of Computer Science, Brigham Young University, Provo, Utah (1995)
15. Liddle, S.W., Embley, D.W., Woodfield, S.N.: Cardinality Constraints in Semantic Data Models. Data and Knowledge Engineering 11(3), 235–270 (1993)
16. Liddle, S.W., Embley, D.W., Woodfield, S.N.: A Seamless Model for Object-Oriented Systems Development. In: Bertino, E., Urban, S. (eds.) ISOOMS 1994. LNCS, vol. 858, pp. 123–131. Springer, Heidelberg (1994)
17. Liddle, S.W., Embley, D.W., Woodfield, S.N.: Unifying Modeling and Programming through an Active, Object-Oriented, Model-Equivalent Programming Language. In: Papazoglou, M.M.P. (ed.) ER 1995 and OOER 1995. LNCS, vol. 1021, pp. 55–64. Springer, Heidelberg (1995)
18. Liddle, S.W., Clyde, S.W., Woodfield, S.N.: A Summary of the ER 1997 Workshop on Behavioral Modeling. In: Chen, P.P., Akoka, J., Kangassalu, H., Thalheim, B. (eds.) Conceptual Modeling. LNCS, vol. 1565, pp. 258–271. Springer, Heidelberg (1999)
19. Liddle, S.W.: Formal Foundations of Conceptual Modeling for Learning from Surprises. In: Position Paper, Workshop on Active Conceptual Modeling of Learning, May 10-12, 2006, Space and Naval Warfare Systems Center, San Diego (2006)
20. Mellor, S.J., Balcer, M.J.: Executable UML: A Foundation for Model-Driven Architecture. Addison-Wesley, Reading (2002)
21. Petri, C.A.: Kommunikation mit Automaten. PhD Dissertation, University of Bonn (1962)

Actively Evolving Conceptual Models for Mini-World and Run-Time Environment Changes

P. Radha Krishna[1] and Kamalakar Karlapalem[2]

[1] Institute for Development and Research in Banking Technology, Hyderabad, India
[2] International Institute of Information Technology, Hyderabad, India
prkrishna@idrbt.ac.in, kamal@iiit.ac.in

Abstract. Run-time application environments are affected by the changes in mini-world or technology changes. Large number of applications are process driven. For robust applications that can evolve over time, there is a need for a methodology that implicitly handles changes at various levels from mini-world to run-time environment through a layers of models and systems. In this paper, we present ER* methodology for evolving applications. In the context of this paper, the role of two-way active behaviour and template driven development of applications is presented. This methodology facilitates capturing active behaviour from run-time transactions and provides a means of using this knowledge to guide subsequent application design and its evolution.

1 Introduction

Due to technology advancements, new market conditions and new laws, organizations need to constantly refine their processes in order to effectively meet the changing requirements. Many of the changes or expected exceptions with respect to evolution of business environment may not be available to model the business processes when they were first envisaged. This is especially true in the case of e-commerce and e-business applications. For example, in an e-contract system, a contract statement "......On receiving the request, Contractor will allocate a CMR (Change Management Request) number to that request and will notify it to the Purchaser. The contractor will then evaluate the need of *this change* with respect to Priority, Feasibility of the change, and Impact on time frame and cost." is not clear to model it unless the actual change is provided to the system. This can happen only during the enactment and it requires changes based on the *new change* request. Thus, modeling a complex application at first instance does not reflect a complete scenario. This calls for an iterative active methodology that constantly *monitors run-time environment and changes in real-world specifications to keep the deployed applications/processes current.*

The conceptual modeling of an application is the first important step towards understanding business process. Entity-Relationship (ER) diagram is a simple but powerful conceptual model that is used to describe the fundamental concepts of a specific domain, their structure and the relationships among them. The basic ER

P.P. Chen and L.Y. Wong (Eds.): ACM-L 2006, LNCS 4512, pp. 57–71, 2007.
© Springer-Verlag Berlin Heidelberg 2007

constructs are entities, relationships and attributes. Usually, a well-defined conceptual model leads to the development of an effective and reliable application. An application system design usually follows developing conceptual model, logical model, physical model, implementation architecture and finally deployment of system. In today's business environments, it is necessary to envisage and understand the operations of business processes and changes that occur during its lifetime. Further, the e-business applications mostly involve multiple organizations and multiple parties. This necessitates the need of active behaviour to synchronize the changes in business logic and business processes across different levels of conceptual/logical models. The concepts/notations used in this paper are as follows:

ER^* meta-model: It is a model about ER^* models used to instantiate a data model for a specific instance of an application.

ER^* model: It is an instance of ER^* meta-model.

ER^* methodology: It is a methodology to design of applications (including conceptual model, logical model and workflow and process models) leading to the deployment of applications.

Template: It is an ER^* meta-model with constraints for a specific application.

Application: Application can range from business processes to regular stand-alone applications and large-scale applications across multiple systems/organizations.

In this paper, we are proposing a methodology that we envision can meet the challenges of evolving conceptual model and the related logical models and run-time environment to rapid changes in mini-world and run-time environment. The manifestation of this methodology to be implemented for a complex real-world application would require appropriate level of effort and human inputs along with design and deployment tools. We scope this paper towards describing this methodology and illustrating it with an example. It is evident that lot of work needs to be done to further develop this methodology for various complex real-world applications.

The rest of the paper is organized as follows: Section 2 describes the need of ER* model. Section 3 introduces ER* methodology and section 4 describes the ER* methodology for evolving applications. Section 5 presents template selection driven evolution for ER models followed by an example in Section 6. Section 7 describes the mechanism for event handling and template selection for evolving applications. Related work is presented in Section 8 and finally we conclude in Section 9.

2   Need for ER* Model

ER models help in modeling a business process and provide visual representation of a database schema for the real-world perception of data, rules, and business processes (in separate but related suite of models), instead of single limited conceptual/logical database schema. In earlier works, several extensions to ER model have been presented by introducing new concepts such as aggregation, rules, and exceptions to support additional requirements of applications. In this work, we use a suite of ER Models, referred as ER* meta-model and its corresponding methodology, which act as a template for modeling the change requirements during application evolution. It models data and process/functional requirements of an application. A template typically has a set of fixed concepts to describe various requirements, specific entities that have specific meaning of entities (such as party and payments), built-in relationships, set of constraints, set of rules, a mapping methodology and consistency requirements. But in the real world, the execution of an application is influenced by various factors, some of which may not be envisaged during the requirements analysis. In order to handle applications in a dynamic environment, an ER* model has to be instantiated by invoking the most appropriate template that best meets the requirements of the changed environment.

Figure 1 shows a macro level description of ER* conceptual modeling framework for evolving applications. There are two major factors that influence the evolution namely, the mini-world and run-time environment. The changes must be appropriately handled for proper execution of the system. Any change in the parameters in the mini-world generates necessary stimuli to initiate actions to modify the ER model. Therefore, ER* model gets adapted by the actions to conform to the changed mini-world.

The run-time environment manages and keeps track of the parameters during its application execution. Sometimes in e-business environment, applications are executed by generating appropriate workflows and executing the tasks/activities therein. During the execution, there may be changes in workflows or exceptions may be raised at run-time. For example, non-delivery of goods by a specified time may lead to delayed payments. Similarly, a network failure (an exception) may prevent an electronic fund transfer in time. Such exceptions or changes have to be handled at run-time. This necessitates that the ER* model has to be adaptable to incorporate such occurrences and provide a remedy for continuing the application execution. In [7], the methodologies to handle such exceptions is presented.

Specifically for e-contract applications the base ER model starts by initiating a template from ER* meta-model and drives the system. The template contains the semantics of the system. Different semantics can give raise to different scenarios of application. Modeling of applications requires both human and system driven specification and deployment in order to handle the active behaviour of applications. A problem of current interest is to manage movement from one template to another at run-time or even evolve the template, if required (see section 6 for an illustrative example).

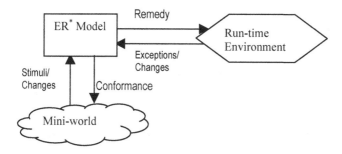

Fig. 1. Conceptual modeling framework for evolving applications

3 ER* Methodology

Our ER* methodology is based on the approach proposed by Batini et al. [1] for designing database applications. Figure 2 shows the application process design model in which the entire design basically divides into two concurrent segments. These two segments are dependent on each other to effectively design the entire application. One segment corresponds to the conceptual design aspects of the application and the other one corresponds to the active behaviour of the application execution. The requirement collection forms the basis for complete process design. Model requirements facilitate the conceptual design where as the process requirements facilitate the execution aspects of the application. As shown in the figure 2, the relationships between the two segments facilitate capture of the dynamic behaviour of applications and incorporate the necessary updates required at the conceptual level of the design process. Here, the ER* model will act as a library of meta-models or templates. This template library was built over knowledge created while working with various applications. That means, it has a set of domain specific templates. The conceptual schema is developed by instantiating an ER model from ER* model. Once a template is selected for a specific application requirement according to the business requirements, the logical schema for both databases and logical level processes is derived after mapping the model components into run-time workflow components.

The static and dynamic behaviour of applications are handled in the right-hand segment. The static criteria can be verified at the application specification and used in the matching procedure with conceptual schema. The events that will raise during the execution along with ECA rules are derived from the application specifications. Expected exceptions are also captured during this step. Exceptions and ECA rules are helpful in specifying the logical level processes. These specifications are used in run-time workflow mapping. The dynamic behaviour will be handled by writing log records from the execution of applications. A knowledge base is built from the unexpected events and exceptions (due to mini-world and business process changes). This knowledge base will form as a source for facilitating active behaviour and incorporates the new rules that manage the exceptions and events. The ER model is modified either by updating the conceptual schema or instantiating a model instance from an appropriate template. In case no suitable template is identified to instantiate the appropriate ER model, a new template needs to be designed and added to ER* meta-model. The process model is also modified according to the application evolution.

The ER* methodology is iterative and is driven by three validation steps, namely, structural validation, functional validation, and behavioral validation. Structural validation is the traditional correctness check of a conceptual model in correctly modeling the data requirements. Functional validation ensures that the applications and the transactions meet the application requirements, and map to the functionality that exists in the mini-world. Behavior validation is the one that is triggered by the constant monitoring of the run-time environment and mini-world changes. This is captured by means of exceptions and user feedback to detect the mismatch between the implemented applications/processes and what exists in the mini-world. In a highly process oriented environments these exceptions can lead to process evolution and

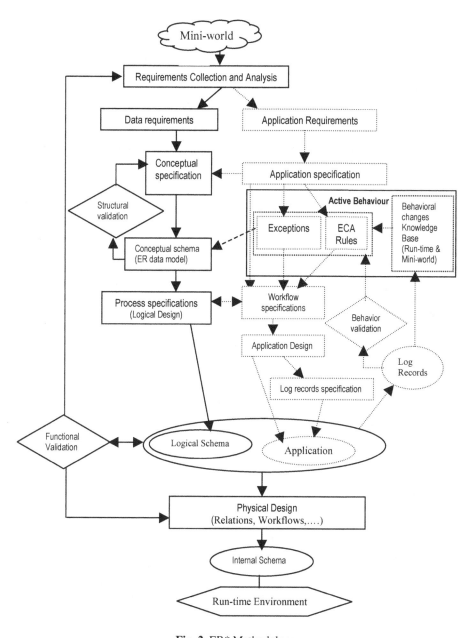

Fig. 2. ER* Methodology

work specification changes. Therefore, these three validation steps are the driver for the ER* methodology for evolution. The log records, exception events, and ECA rules are part of the driver for supporting this methodology.

ECA rules facilitate monitoring and execution of an application. The ECA rules are well established in the context of databases, which has the functionality of active capability that satisfies the needs of various applications and timely response to a situation. Initially, application execution behaviour (and their underlying workflows), and events as well as snapshots of run-time environment, will be fed into an active database [13, 15]. In order to manage the active behaviour of applications effectively, the semantic behaviour of rules and exceptions must be captured. It can be modeled by considering the events and actions along with the update primitives for conceptual modeling the application evolution. These in turn are held in learning appropriate changes to conceptual models (or templates).

4 ER* Methodology for Evolving Applications

Figure 3 shows the ER* methodology for evolving applications. The main idea is to facilitate deployment of data models that sustain over a period of time. This approach follows a two-way perspective of actively evolving conceptual models: (i) across the time domain (present, past and future) and (ii) at various levels (meta, conceptual, logical and application level). The main component is the active behaviour, which learns the execution behaviour and, accordingly makes the changes required at various levels. Further, it also propagates the changes to the next generation (from past to present to future). Each vertical segment of figure 3 follows the ER* methodology described in the section 3.

The evolution from present to future can be handled in one of the following three ways based on the available tools that support the evolution:

- Template selection
- Operator assisted evolution of ER models
- Complete re-design of ER models (from scratch)

The template selection mechanism manifests itself as a ER^* methodology problem where in an evolving application needs to select appropriate template to meet its application requirements. A specific case of e-contract is presented in the next section to illustrate one such environment. In the second case, a set of operations have to be developed in order to handle the active behaviour so that these can be programmed to (semi-) automatically (with human intervention) design/update/derive the suitable ER models. The last approach requires an automated system that learns the behaviour and arrives at designing appropriate ER models. This requires a sophisticated set of tools, and to the best of our knowledge, no current tools are available to automate this completely. Currently, we are in the process of developing taxonomy for the operations to carry out ER models according to the evolution of applications.

The ER* methodology is helpful for maintenance, archiving and retrieval, besides providing a library of ER* meta-models specific to various domains. This approach is also useful when an application system is located in various countries. A typical standard application such as core banking solution (CBS) in a bank also requires

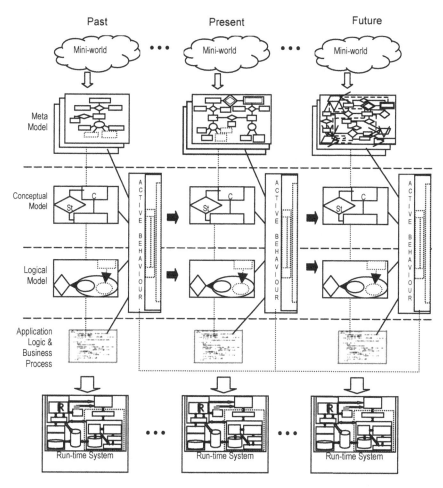

Fig. 3. ER* methodology for evolving information systems

evolution when it is placed in different countries due to international/local rules and regulations. An application needs to evolve whenever these rules and regulations change. Thus, *there is a notion of spatio-temporal dimension of evolving application and matching them to different conceptual models.*

5   Template Selection Driven Evolution

In this section, we discuss on several possibilities that may arise during application execution which may eventually require a change in the ER model. In order to visualize or incorporate such revisions in the conceptual model, the ER* model has to adapt appropriately. Below we discuss three possible ways to deal with such a dynamic scenario.

a) Whenever there is a change in the execution, an appropriate ER model can be instantiated from ER* model (figure 4) and necessary modifications can be made on it depending on the revised scenario. Such an approach is suitable only when there is a change in the number of entities involved. Here, there is no major structural change in the model.

b) The second possibility could be that an application requires one or more additional template elements, which are to be incorporated in the ER model. This can possibly be handled by maintaining a set of ER models corresponding to several possible application scenarios and then instantiating a template from the most appropriate ER model (figure 5). In the event of a change, one can move from an instance of one template to an instance of another template which can further drive the application execution.

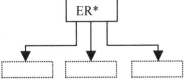

Fig. 4. ER* Model Instantiation

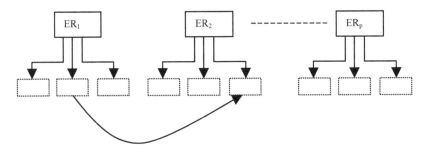

Fig. 5. Template instantiation from multiple ER models

c) The change could evolve the template itself. By this we mean that one can always use a standard ER model and instantiate a template when an e-application is executed but as time progresses, to cope up with the changing scenario, evolve the template by adding or modifying the schema concepts. However, this approach is difficult to implement, as it requires complete structural changes within a template.

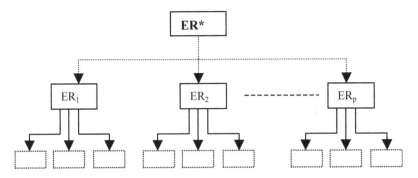

Fig. 6. ER* model

We combine the first two approaches and introduce the concept of a Global ER model (figure 6), referred as ER*. This ER* model can be thought of as union of all ER models which can deal with different possible scenarios. As and when situation demands, due to changes in the run-time environment or mini-world, an appropriate model(s) can be instantiated in order to seamlessly model the data and processes. This way of adapting an ER* model can be visualized as having a global template comprising of all the concepts such as entities, attributes, relations, and aggregations to act as template elements that are required in various possible execution scenarios. To start with, at the beginning of execution, the most appropriate template elements can be activated and as time progresses in the event of any change some of the template elements can be deactivated and some other can be activated to meet the requirement. This gives rise to a new instance of the template that is used in the changed context. New template elements are added in case there is no suitable concept available. That is, template elements are added or modified based on the active behaviour of business transactions during their execution due to run-time changes or mini-world changes.

6   ER* Example of E-Contract Application

In this section, we provide an e-contract example. Figure 7 shows EREC meta-model for an e-contract [10]. This meta-model serves as a template. A detailed description of

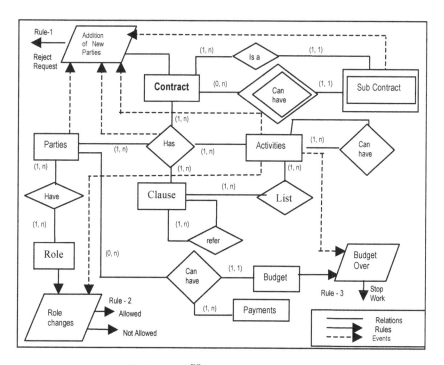

Fig .7. An ER^{EC} Meta Model for E-Contract

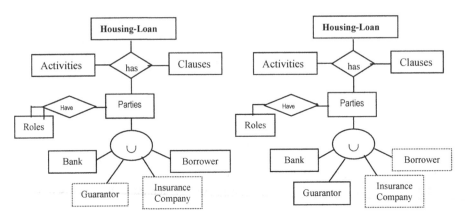

Fig. 8. Standard template of Housing-Loan Contract

Fig. 9. Template with change of roles

this model can be found in [10, 15]. Consider a "housing-loan" contract between a borrower and the bank. According to the contract the monthly repayment for the loan is to be made by borrower. The contract is enacted as per a standard template where the parties are, the bank, borrower, guarantor, insurance company (figure 8). However, at the beginning, the bank and the borrower are only the active parties and the guarantor, insurance company do not have any active role to play in the contract. For simplicity, the diagrams used to depict the templates are shown only with few elements that are necessary for understanding.

Case 1: (Run-time change) - Borrower defaults
As per the contract the borrower is supposed to repay the monthly installment by a due date. In case the borrower becomes defaulter for more than a specified period, then the bank contacts the guarantor to repay the installments. Here, the role of the guarantor has changed to borrower. Thus, a new template is instantiated by modifying the role of the party 'Guarantor'. The template shown in the figure 9 depicts these changes.

Case 2: (Run-time change) - Borrower's death/disablement
In the event of death or any disability of the borrower, a new sub-contract between the bank and insurance company starts enactment. This sub-contract may require a different set of parties, activities, clauses, etc. Further, it may have few sub-sub-contracts to meet the requirements such as police verification, medical certificate etc. Now, the original template has to be replaced or augmented with an instance of a template of another ER schema. Here, the two possibilities are (i) the insurance company pays the balance amount in the case of borrower expired and the house is allotted to the nominee; (ii) the insurance company releases a compensation amount in the case of borrower has disablement. Figure 10 shows a sub-contract, "insurance-claim" to instantiate a new template. Note that there was no sub-contract entity in the original template (figure 8).

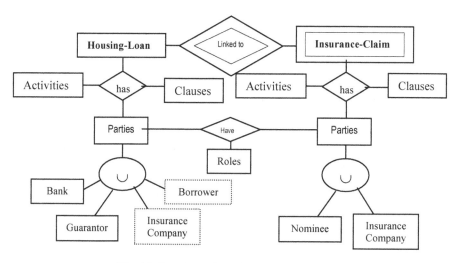

Fig. 10. Template with addition of sub-contract

Case 3: (Mini-world change) –

Road expansion
Consider a situation, say a house built with the loan amount has to be dismantled due to expansion of a road. In this case, the government has to compensate a prescribed amount to the borrower for the portion of the house that was damaged. Here, not only adding a party (i.e., government) takes place, but also the policies that govern different activities changes. This scenario might not have been conceived during requirement analysis. In order to adapt this mini-world change, it would require creating a suitable template by augmenting new concepts and/or modifying existing template elements. In the housing-loan contract, the new concepts could be virtual entities – society, human rights, besides adding a party like government, which have to be modeled appropriately in the template (figure 11). In the figure the virtual-entity is shown as a dashed rectangle.

In order to handle mini-world and run-time changes in the contract execution environment, appropriate templates must be selected. Usually, when a contract begins a standard template is chosen from the repository to drive the contract. As the contract evolves specific templates can be arrived at to suit a particular application under consideration. However, if standard templates are not available to deal with certain unforeseen events, then a generalized template can be used to drive the contract instead of an abnormal termination of the contract. The generalized template elements are actually a superset of standard template elements and human intervention is necessary to choose suitable template elements to be considered in the generalized template. Thus, starting with a standard template the template selection process can move in either direction to a more generalized template or a specific one.

This example shows the complexities involved in coming up with ER* meta-models and the amount of human intervention and tools that required semi-automate support of evolving applications.

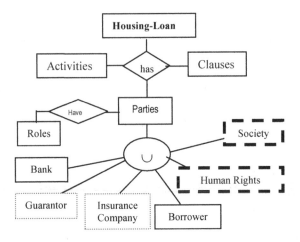

Fig. 11. Template with additional concepts

7 Event Handling and Template Selection Mechanism for Evolving Applications

In this section, we describe the handling of events and selection of templates from templates repository at run-time. We consider the changes in the mini-world and/or runtime environment as events that would require appropriate action for the application execution. For example, an event may be caused due to change in the policy or an exception in the run-time environment. This may affect the entities in the ER* model. Thus, the event handler plays a major role in the application evolution. Depending on the event, the event handler generates new instance of ER* meta-model, if required, by invoking appropriate template.

Application evolution requires the modification of schema definitions and in-turn changes in workflow definitions, as a remedy. Moreover, additional ECA constructs are needed in the system during work in progress; an advanced schema evolution capability is required at run-time. The meta-modeling approach to e-contract modeling facilitates application evolution and thereby workflow enactment. The present work offers a practical solution from application modeling, enactment, to evolution. We maintain application evolution policies that refer to adaptable instances of ER* schema while the application is executing. We enhanced the capabilities of ADOME-WFMS [8, 9] to facilitate the application evolution with the help of schema evolution and generate *evolution patterns* to affect changes in an ER* model. Evolution patters are specific to applications under consideration. For example, in an e-contract application, the evolution patterns could be of the form {e-contract}$^+${Activity}$^+${Clause}*{Action}*. These patterns are useful to describe how the evolution changes will be specified, implemented and perceived.

Figure 12 shows an architecture of a system to actively model ER* meta-model and its instances. The run-time environment details such as workflows, rules, etc. are maintained in the database. *Workflow Generation / Specification Subsystem* generates workflows and rules. It also allows the administrators to customize and edit them.

Workflow definitions created or specified are executed by the *E-ADOME workflow engine*. That is, the workflow engine enacts the workflows specified by the workflow Generation/specification subsystem.

The *Event Handler* manages the events occurring during the execution of workflows. It handles events in a unified manner for both normal and exception parts of a business process workflow. The *ECA Rule Manager* generates appropriate ECA rules based on the input from Event Handler. It also keeps track of generated rules with their corresponding actions and allows users to define additional rules if necessary.

The workflow engine and the ECA rule manager works in a synchronized manner. Thus, the ECA rules control the workflow execution and the events that occur during the workflow execution result in appropriate actions. The changes in the mini-world update the corresponding database. The mini-world database maintains the currency of information such as evolution policies, versions etc. that governs execution of the application. *Metadata Database* captures the updates that taken place in the run-time as well as in the mini-world. These updates become input to the *Run-time Template Evaluator*. Run time template evaluator generates candidate templates and add/modify the template repository.

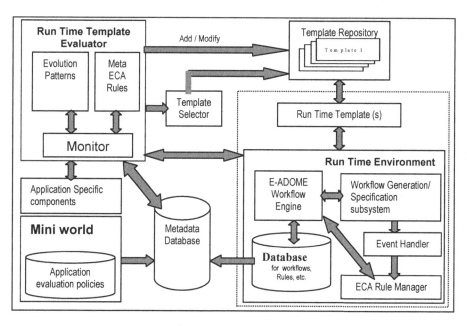

Fig. 12. An ER* architecture for evolving applications

Template Repository contains the templates that are specific to application under consideration in XML format. Templates are added or modified based on the requirements collected from runtime environment changes as well as mini-world changes. Here, *Monitor* acts as broker between the run time environment and mini-world. It generates/modifies the *evolution patterns* and *(meta-) ECA rules*. Monitor receives events and results from the run time environment and mini-world, as well as from application specific components. The modeling of changes in the application evolution can be seen as a different kind of meta processes (tasks). These meta processes can be modeled and evolution patterns and meta ECA rules can be extracted from the meta processes. Further, *Application Specific Components* are required for encapsulation and realization of the domain-specific logic for the application. The three components namely Monitor, evolution patterns and meta ECA rules will serve as Run-time template evaluator. Template selector selects the appropriate Run Time template(s) from the Template Repository, which in turn drives the application execution.

8   Related Work

The ER model, introduced by Chen [4] is a standard conceptual model used in the design of relational databases. In the literature, there are several extensions to original ER model. Tanaka et al [16] developed a conceptual modeling framework for rule processing and ECA triggers in active databases. An approach to schema evolution is presented in [6]. They developed EVER (evolutionary ER) diagrams to specify schema versions, relationships among attributes and the conditions for maintaining consistent views of programs.

The ER^{EC} model presented in [10] provides a meta-model for describing the data, relationships and the associated information such as rules and exceptions to support modeling electronic contracts, and the e-contract meta-model serves to build a model of models for e-contracts. Chen et al [5] presented several important research directions in conceptual modeling. Out of eight areas proposed by them, active modeling is one of the future research directions. It was stressed that data changes and schema changes in the real-world and at any given time, a conceptual model can be as a multi-level and multi-perspective abstraction of reality. This direction motivated us to develop evolving conceptual models.

Casati et al. [2] provided workflow evolution, from both a static and dynamic point of view. Proper and Weide [14] described a theory of evolving application models and introduced the concepts relating schema evolution that are independent of modeling techniques. Kersten and Szpakowicz [12] modeled business negotiations for electronic commerce using software agents. An ER^{EC} model is presented for conceptual modeling electronic contracts and presented a framework for their enactment [15]. Various modeling frameworks and deployment scenarios of e-contracts are presented in [11].

9   Conclusion

In this paper, we proposed a methodology of actively evolving ER^{*} meta-models that can cater to various application domains. A specific ER^{*} model handles a specific application requirement scenario for a specific duration of time. Whenever mini-world and run-time environment changes, the corresponding data models at various levels and applications themselves needs to be changed. The idea presented in this paper is useful in visualizing this evolution procedure and develop specific methodologies, procedures and tools to actively support application evolution. A key ingredient in our methodology is the use of monitoring and bookkeeping of mini-world and run-time changes to propose an ECA rule based solution.

References

1. Batini, C., Ceri, S., Navathe, S.B.: Conceptual Database Design - An Entity-Relationship Approach. The Benajamin/Cummings Publishing Company Inc., Redwood City, California (1992)
2. Casati, F., Ceri, S., Pernici, B., Pozzi, G.: Workflow Evolution. In: Thalheim, B. (ed.) ER 1996. LNCS, vol. 1157, Springer, Heidelberg (1996)
3. Chakravarthy, S., Krishnaprasad, V., Anwar, E., Kim, S.-K.: Composite events for active databases: Semantics, contexts and detection. In: Proceedings of the 20th VLDB conference, Santiago, Chile, pp. 606–617 (1994)
4. Chen, P.P.: The entity-relationship model - towards a unified view of data. ACM Transactions on Database Systems 1(1) (1976)
5. Chen, P.P., Thalheim, B., Wong, L.Y.: Future Directions of Conceptual Modeling. In: Chen, P.P., Akoka, J., Kangassalu, H., Thalheim, B. (eds.) Conceptual Modeling. LNCS, vol. 1565, Springer, Heidelberg (1999)
6. Chien-Tsai, L., Chrysanthis, P.K., Chang, S.-K.: Database Schema Evolution through the Specification and Maintenance of Changes on Entities and Relationships. In: Loucopoulos, P. (ed.) ER 1994. LNCS, vol. 881, pp. 132–149. Springer, Heidelberg (1994)
7. Chiu, D.K.W., Li, Q., Karlapalem, K.: A meta-modeling approach for workflow management system supporting exception handling. Information Systems 24(2), 159–184 (1999)
8. Chiu, D.K.W., Karlapalem, K., Li, Q., Eleanna, K.: Workflow View Based E-Contracts in a Cross-Organizational E-Services Environment. Distributed and Parallel Databases 12(2/3), 193–216 (2002)
9. Chiu, D.K.W., Li, Q., Karlapalem, K.: Web interface driven cooperative exception handling in ADOME workflow management system. Information Systems 26(2), 93–120 (2001)
10. Karlapalem, K., Dani, A.R., Krishna, P.R.: A Frame Work for Modeling Electronic Contracts. In: Kunii, H.S., Jajodia, S., Sølvberg, A. (eds.) ER 2001. LNCS, vol. 2224, Springer, Heidelberg (2001)
11. Karlapalem, K., Krishna, P.R.: State-of-the-Art in Modeling and Development of Electronic contracts. In: Roddick, J.F., Benjamins, V.R., Si-Saïd Cherfi, S., Chiang, R., Claramunt, C., Elmasri, R., Grandi, F., Han, H., Hepp, M., Lytras, M., Mišić, V.B., Poels, G., Song, I.-Y., Trujillo, J., Vangenot, C. (eds.) Advances in Conceptual Modeling - Theory and Practice. LNCS, vol. 4231, pp. 3–4. Springer, Heidelberg (2006)
12. Kersten, G.E., Szpakowicz, S.: Modeling business negotiations for electronic commerce. In: Proceedings of the 7th workshop on Intelligent Information Systems, IPI PAN, Warsaw, pp. 17–28 (1998)
13. Morgenstein, M.: Active databases as a paradigm for enhanced computing environments. In: Proceedings of the International Conference on Very Large Databases (1983)
14. Proper, H., Van der Weide, T.P.: A General Theogy for Evolving Application Models. IEEE TKDE (1995)
15. Krishna, P.R., Kamalakar, K., Chiu, D.K.W.: An EREC Framework for E-Contract Modeling, Enactment and Monitoring. Data and Knowledge Engineering 51(1), 31–58 (2004)
16. Tanaka, K.S., Navathe, B., Chakravarthy, S., Karlapalem, K.: ER-R: An enhanced ER model with situation-action rules to capture application semantics. In: ER 1991, pp. 59–75 (1991)

Achievements and Problems of Conceptual Modelling

Bernhard Thalheim

Department of Computer Science,
Christian Albrechts University Kiel, 24098 Kiel, Germany
thalheim@is.informatik.uni-kiel.de

Abstract. Database and information systems technology has substantially changed. Nowadays, content management systems, (information-intensive) web services, collaborating systems, internet databases, OLAP databases etc. have become buzzwords. At the same time, object-relational technology has gained the maturity for being widely applied. Conceptual modelling has not (yet) covered all these novel topics. It has been concentrated for more than two decades around specification of structures. Meanwhile, functionality, interactivity and distribution must be included into conceptual modelling of information systems. Also, some of the open problems that have been already discussed in 1987 [15,16] still remain to be open. At the same time, novel models such as object-relational models or XML-based models have been developed. They did not overcome all the problems but have been sharpening and extending the variety of open problems. The open problem presented are given for classical areas of database research, i.e., structuring and functionality. The entire are of distribution and interaction is currently an area of very intensive research.

The presentation of open problems is combined with the introduction to the achievements of conceptual modelling. The paper develops an approach to conceptual modelling for object-relational, collaborating information systems that support virtual communities of work, integration of information systems, varieties of architecture such as the OLTP-OLAP architecture, varieties of play-out and play-in systems, and data analysis engines. The paper is based on an extended entity-relationship model that covers all structuring facilities of object-relational systems. It uses the theory of media types and storyboards for the specification of interactivity and provides a framework for collaboration.

The paper presents 20 open problems that need to be solved for conceptual modelling. The problems are sketched. Main references and the background are given. Additional references can be provided by the author on demand.

1 Introduction

1.1 Information Systems Design and Development

The problem of information system[1] design can be stated as follows:

Design the logical and physical structure of an information system in a given database management system (or for a database paradigm), so that it contains

---

[1] A database system consists of a number of databases and a database management system. An information system extends the database system by the application system and by presentation systems.

P.P. Chen and L.Y. Wong (Eds.): ACM-L 2006, LNCS 4512, pp. 72–96, 2007.
© Springer-Verlag Berlin Heidelberg 2007

all the information required by the user and required for the efficient behavior of the whole information system for all users. Furthermore, specify the database application processes and the user interaction.

The implicit goals of database design are:

- to meet all the information (contextual) requirements of the entire spectrum of users in a given application area;
- to provide a "natural" and easy-to-understand structuring of the information content;
- to preserve the designers entire semantic information for a later redesign;
- to achieve all the processing requirements and also a high degree of efficiency in processing;
- to achieve logical independence of query and transaction formulation on this level;
- to provide a simple and easily to comprehend user interface family.

Over the last years database structures have extensively been discussed. Some of the open questions have been satisfactorily solved. Modelling includes, however, more aspects:

Structuring of a database application is concerned with representing the database structure and the corresponding static integrity constraints.

Functionality of a database application is specified on the basis of processes and dynamic integrity constraints.

Distribution of information system components is specified through explicit specification of distribution.

Interactivity is provided by the system on the basis of foreseen stories for a number of envisioned actors and is based on media objects which are used to deliver the content of the database to users or to receive new content.

This understanding has led to the **co-design approach** to modelling by specification **structuring**, **functionality**, **distribution**, and **interactivity**. These four aspects of modelling have both syntactic and semantic elements.

Nevertheless, the main open problem has not yet been solved:

Open problem 1.

Find a common motivation, a common formal model and a correspondence that justify the properties and formalize the characteristics.

1.2   Information System Models in General

Database design is based on one or more database models. Often, design is restricted to structural aspects. Static semantics which is based on static integrity constraints is sparsely used. Processes are then specified after implementing structures. Behavior of processes can be specified by dynamic integrity constraints. In a late stage, interfaces are developed. Due to this orientation the depth of the theoretical basis is different as shown in the following table displaying the state of the art in the 90ies:

	Used in practice	Theoretical back-ground	Earliest layer of specification
Structures	well done	well developed	strategic
Static semantics	partially used	well developed	conceptual
Processes	somehow done	parts and pieces	requirements
Dynamic semantics	some parts	parts and glimpses	*implementation*
Interfaces	intuitive	nothing	*implementation*
Stories	intuitive	nothing	*implementation*

Database design requires consistent and well-integrated development of structures, processes, distribution, *and* interfaces. We will demonstrate below that extended entity-relationship models allow to handle all four aspects.

Database systems are now extended to web information systems, to data warehouse, to intelligent knowledge bases, and to data analysis systems. This extension can be developed in a conservative fashion or based on novel paradigms. As long as novel paradigms do not overcome the problematic parts of database systems operating, conservative extension must be preferred. In this case we need a good architecture [9,7] for extensions of systems.

Open problem 2.

Find an architecture for generic extension of database systems that entirely supports the extensions, that allows to model all facilities and that supports reasoning on system properties.

At the same time, we need to model database systems quality. The quality criteria are often stated in a rather fuzzy form. Typical quality criteria are [4] accuracy, changeability, fault tolerance, operability, performance, privacy, recoverability, reliability, resource efficiency, safety, security, stability, and testability.

Open problem 3.

Define in a formal form quality criteria, quality attributes, and quality metrics for conceptual modelling and develop a framework for their enforcement, control, and refinement.

2   Towards a Science of Modelling

2.1   The Documentation and Knowledge Gap in Information Systems Modelling

Managing information systems applications becomes crucial for evolution and prosperity of companies. Information systems are used over decades whereas software and hardware changes more often. For this reason, the complete information systems development process must be well-documented. The entire development knowledge that is necessary for building, using, and maintaining information systems must be kept over the lifetime of the system. The documentation includes meta-data on the enterprise system, the technical process supporting the system, the business processes supported by

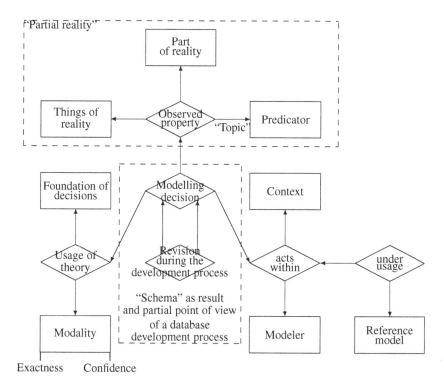

Fig. 1. The Knowledge Gap on Database Design Decisions

the system, services and software tools used within the system, organizational policies and people.

Classical information systems development follows often the late documentation approach, i.e. the development documentation is based on the final result of the conceptualization, i.e. consists of the conceptual schema and its translation to relational or object-relational system languages. Already [5] discussed the forgetful development of software products. Classically we observe that

- developers base their design decisions on a "partial reality", i.e. on a number of observed properties within a part of the application,
- developers are developing the information system within a certain context,
- developers reuse their experience gained in former projects and solutions known for their reference models, and
- developers use a number of theories with a certain exactness and rigidity.

The design decisions made during the design process are deeply influenced by these four hidden factors. In some approaches revisions made during the information systems development are recorded. However, since the background knowledge is not recorded the documentation of the information systems development is fragmentary[2]. This knowledge gap is visualized in Figure 1 [18].

The most pressing challenges that organizations are currently facing are: provide IT portfolio management, reduce IT redundancy, prevent IT applications failure, reduce IT expenditures, enable knowledge management, adhere to regulatory requirements, and enable enterprise-wide integration of applications.

Open problem 4.

Develop a specification language that allows to specify the modelling decision together with the causes for choosing the given decision and that allows to reason on consistency of causes.

2.2   Modelling in General and for Information Systems

Conceptual modelling and modelling differ a lot. Mathematical modelling satisfies three requirements based on a chosen modelling language[21]:

- adequate and approximate representation $\mathcal{R}$ of phenomena $\mathcal{P}$ observed in reality;
- unique representation of these phenomena without any choice for a representation style;
- repetitiveness of the result of modelling whenever modelling is repeated.

The process of modelling thus includes the following steps:

1. formulation of laws, general properties, associations etc. within the chosen modelling language and statement of problems and requirements under consideration;
2. solution of the problems and development of a theory for the solution based on methods developed so far;
3. check whether the models satisfies all requirements of practical exploitation and development of refined models;
4. development of theories on the practical application based on the model.

Conceptual modelling considers mainly the first process step. The second step is often neglected. The third step comes into play whenever practical exploitation leads to a

---

[2] Due to our involvement into the development and the service for the CASE workbenchs (DB)[2] and ID[2] we have collected a large number of real life applications. Some of them have been really large or very large, i.e., consisting of more than 1.000 attribute, entity and relationship types. The largest schema in our database schema library contains of more than 19.000 entity and relationship types and more than 60.000 attribute types that needs to be considered as different. Another large database schema is the SAP R/3 schema. It has been analyzed in 1999 by a SAP group headed by the third author during his sabbatical at SAP. At that time, the R/3 database used more than 16.500 relation types, more than 35.000 views and more than 150.000 functions. The number of attributes has been estimated by 40.000. Meanwhile, more than 21.000 relation types are used. The schema has a large number of redundant types which redundancy is only partially maintained. The SAP R/3 is a very typical example of a poorly documented system. Most of the design decisions are now forgotten. The high type redundancy is mainly caused by the incomplete knowledge on the schema that has been developed in different departments of SAP.

change of the model itself. According to observations for application developed with the workbenchs $(DB)^2$ and ID^2 this third step starts about half a year after introduction of of the solution to practice.

Open problem 5.

Develop a theory of schema evolution that allows to change the scope of the model, i.e. the partial reality, and that allows to reason on the impact of these changes.

According to [14], a theory consists of (1) an abstract calculus and a set of postulates and (2) a set of rules that assign an empirical content to the calculus and the postulates by providing 'coordinating definitions' and 'empirical interpretations'. Conceptual modelling typically results in the first part and assumes that the second part is intuitively understood by the user of the model.

[5] extracted three main properties of model relationships $R(\mathcal{P}, \mathcal{R}, \mathcal{A})$ between phenomena, representations and actors $\mathcal{A}$ developing the model:

Mapping property: The modelling relationship is based on a mapping from $\mathcal{P}$ to $\mathcal{R}$ that is a surjective function.

Truncation property: The mapping is based on an explicit abstraction, i.e. does not distinguish all possible properties of phenomena but only those that serve the purpose of the mapping. The properties considered (surplus properties) and the ignored ones are clearly separated.

Pragmatic property: The modelling relationship is based on the scope of the actors, their intentions and tasks, their culture, skills and knowledge, the purposes of modelling and is based on admissible tools and techniques of investigation.

Open problem 6.

Develop a theory of conceptual modelling that allows to reason on the mapping, truncation and pragmatic properties, and their change. This theory must be based on explicit criteria for choosing a certain schema.

2.3 Modelling By Suites of Models

It is well-known [16] that each schema has a large variety of equivalent schemata. Whether we choose the current one depends

- on the designer knowledge, habits, recency, vividness, and congruity,
- on the underlying technology and the corresponding requirements to avoid certain schemata or to prefer certain schemata, and
- on the application and functional requirements especially performance requirements.

All these reasons for preferring one schema over all others may change. If a change occurs we may switch to another schema or we evolve the current schema to a more appropriate one. The current schema is thus a snapshot within a schema suite. Some of the schemata are equivalent to other, are refinements or extensions of other schemata.

A *schema suite* consists of a set of schemata, an integration or association schema among these schemata and obligations requiring maintenance of the association. We specify schemata based on a type system enabling in describing structuring of schemata and functionality of schema changes, in describing their associations through relationship types and constraints. The functionality of schemata changes is specified by a retrieval expression, the maintenance policy and a set of functions supporting the utilization of the schema and the schema suite.

Open problem 7.

Develop a theory of isomorphisms, refinement and extension of conceptual schemata.

Databases are the kernel of information systems. We use databases for extraction of data, for creation of data that have a meaning for the application (content), for extraction of data that can be understood by a user (data-warehouse content), for enhancement of data with other data characterising the data (topic-enhanced data), for protection of data against misuse (protected data), for relating data to certain users (private data), for structuring data in a way that can easily be memorised (meme data), etc. Often the generation of such data can be layered in a form similar to the one in Figure 2.

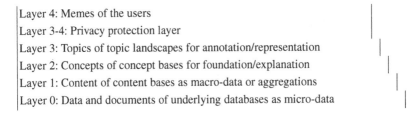

Layer 4: Memes of the users
Layer 3-4: Privacy protection layer
Layer 3: Topics of topic landscapes for annotation/representation
Layer 2: Concepts of concept bases for foundation/explanation
Layer 1: Content of content bases as macro-data or aggregations
Layer 0: Data and documents of underlying databases as micro-data

Fig. 2. An Architecture of Content Management User-Oriented Systems[19]

Each extraction, transformation and loading of data to another database changes also semantics of data, applicability of functions, and pragmatics of data. This change of semantics is well-known for views. Constraints valid in the database used for extraction must not remain to be valid. Other constraints may become valid due to the extraction. Schemata used for loading often use implicit constraints. The same observation can be made for changes in pragmatics, i.e. in the meaning of data for users. Additionally, the operations used for extraction may lead to different meaning or are creating non-sense. This observation can already been made for data warehouses and OLAP databases[6]. A similar change is imposed whenever a database becomes changed due to crucial or inevitable surprises.

Open problem 8.

Develop architectures of layered information systems with appropriate maintenance mechanisms for data at all layers and with transformation of semantics and pragmatics together with extraction and transformation of data.

3   Specification of Structuring

3.1   Languages for Structure Specification

Structuring of databases is based on three interleaved and dependent parts:

Syntactics: Inductive specification of structures uses a set of base types, a collection of constructors and an theory of construction limiting the application of constructors by rules or by formulas in deontic logics. In most cases, the theory may be dismissed. Structural recursion is the main specification vehicle.

Semantics: Specification of admissible databases on the basis of static integrity constraints describes those database states which are considered to be legal. If structural recursion is used then a variant of hierarchical first-order predicate logics may be used for description of integrity constraints.

Pragmatics: Description of context and intension is based either on explicit reference to the enterprise model, to enterprise tasks, to enterprise policy, and environments or on intensional logics used for relating the interpretation and meaning to users depending on time, location, and common sense.

The inductive specification of structuring is based on base types and type constructors. A base type is an algebraic structure $B = (Dom(B), Op(B), Pred(B))$ with a name, a set of values in a domain, a set of operations and a set of predicates. A class B^C on the base type is a collection of elements form $dom(B)$. Usually, B^C is required to be set. It can be a list, a multi-set, a tree etc. Classes may be changed by applying operations. Elements of a class may be classified by the predicates.

A *type constructor* is a function from types to a new type. The constructor can be supplemented with a *selector* for retrieval (such as *Select*) and *update functions* (such as *Insert, Delete*, and *Update*) for value mapping from the new type to the component types or to the new type, with correctness criteria and rules for validation, with default rules, with one or more user representations, and with a physical representation or properties of the physical representation.

Typical constructors used for database definition are the *set, tuple, list* and *multi-set* constructors. For instance, the set type is based on another type and uses algebra of operations such as union, intersection and complement. The retrieval function can be viewed in a straightforward manner as having a predicate parameter. The update functions such as *Insert, Delete* are defined as expressions of the set algebra. The user representation is using the braces $\{,\}$. The type constructors define type systems on basic data schemes, i.e. a collection of constructed data sets. In some database models, the type constructors are based on pointer semantics.

Other useful modelling constructs are *naming* and *referencing*. Each concept type and each concept class has a name. These names can be used for the definition of further types or referenced within a type definition. Often structures include also *optional* components. Optional components and references must be used with highest care since otherwise truly hyper-sophisticated logics such as topoi are required [10]. A better approach to database modelling is the requirement of weak value-identifiability of all database objects [11].

3.2 Integrity Constraints

Integrity constraints are used to separate "good" states or sequences of states of a database system from those which are not intended. They are used for specification of semantics of both structures and processes. Therefore, consistency of database applications can not be treated without constraints. At the same time, constraints are given by users at various levels of abstraction, with a variety of vagueness and intensions behind and on the basis of different languages. For treatment and practical use, however, constraints must be specified in a clear and unequivocal form and language. In this case, we may translate these constraints to internal system procedures which are supporting consistency enforcement.

Each structure is also based on a set of **implicit model-inherent integrity constraints**:

Component-construction constraints are based on existence, cardinality and inclusion of components. These constraints must be considered in the translation and implication process.

Identification constraints are implicitly used for the set constructor. Each object either does not belong to a set or belongs only once to the set. Sets are based on simple generic functions. The identification property may be, however, only representable through automorphism groups [1]. We shall later see that value-representability or weak-value representability lead to controllable structuring.

Acyclicity and finiteness of structuring supports axiomatisation and definition of the algebra. It must, however, be explicitly specified. Constraints such as cardinality constraints may be based on potential infinite cycles.

Superficial structuring leads to representation of constraints through structures. In this case, implication of constraints is difficult to characterize.

Implicit model-inherent constraints belong to the performance and maintenance traps.

Integrity constraints can be specified based on the B(eeri-)V(ardi)-frame, i.e. by an implication with a formula for premises and a formula for the implication. BV-constraints do not lead to rigid limitation of expressibility. If structuring is hierarchic then BV-constraints can be specified within the first-order predicate logic. We may introduce a variety of different classes of integrity constraints defined:

Equality-generating constraints allow to generate for a set of objects from one class or from several classes equalities among these objects or components of these objects.

Object-generating constraints require the existence of another object set for a set of objects satisfying the premises.

A class $\mathcal{C}$ of integrity constraints is called *Hilbert-implication-closed* if it can be axiomatised by a finite set of bounded derivation rules and a finite set of axioms. It is well-known that the set of join dependencies is not Hilbert-implication-closed for relational structuring. However, an axiomatisation exists with an unbounded rule, i.e. a rule with potentially infinite premises.

The main deficiency is the constraint acquisition problem. Since we need a treatment for sets a more sophisticated reasoning theory is required. One good candidate is visual or graphical reasoning that goes far beyond logical reasoning [3].

Open problem 9.

Provide a reasoning facility for treatment of sets of constraints.
Classify 'real life' constraint sets which can be easily maintained and specified.

Additional problems for dependencies can already be stated on the level of the relational model. We conclude this subsection with a number of problems.

Open problem 10.

Is the implication problem for closure dependencies and functional dependencies decidable? Axiomatizable?
Which subclass of inclusion constraints properly containing the unary inclusion dependencies is axiomatizable together with the class of functional dependencies?
Which subclass of join dependencies properly containing the class of multi-valued dependencies is axiomatizable?
Characterize relations which are compatible under functional dependencies.
Characterize the properties of constraint classes under horizontal decomposition.

3.3 Representation Alternatives

The classical approach to database objects is to store an object based on strong typing. Each real life thing is thus represented by a number of objects which are either coupled by the object identifier or supported by specific maintenance procedures. In general, however, we might consider two different approaches to representation of objects:

Class-wise, identification-based representation: Things of reality may be represented by several objects. The *object identifier* (OID) supports identification without representing the complex real-life identification. Objects can be elements of several classes. In the early days of object-orientation it has been assumed that objects belong to one and only one class. This assumption has led to a number of migration problems which have not got any satisfying solution.
Structuring based on extended ER models [16] or object-oriented database systems uses this option. Technology of relational and object-relational database systems is based on this representation alternative.

Object-wise representation: Graph-based models which have been developed in order to simplify the object-oriented approaches [1] display objects by their sub-graphs, i.e. by the set of nodes associated to a certain object and the corresponding edges. This representation corresponds to the representation used in standardization.
XML is based on object-wise representation. It allows to use null values without notification. If a value for an object does not exist, is not known, is not applicable or cannot be obtained etc. the XML schema does not use the tag corresponding to the attribute or the component. Classes are hidden.

Object-wise representation has a high redundancy which must be maintained by the system thus decreasing performance to a significant extent. Beside the performance problems such systems also suffer from low scalability and bad utilization of resources. The operating of such systems leads to lock avalanches. Any modification of data requires a recursive lock of related objects.

For these reasons, objects-wise representation is applicable only under a number of restrictions:

- The application is stable and the data structures and the supporting basic functions necessary for the application are not changed during the lifespan of the system.
- The data set is almost free of updates. Updates, insertions and deletions of data are only allowed in well-defined restricted 'zones' of the database.

A typical application area for object-wise storage are archiving systems, information presentation systems, and content management systems. They use an update system underneath. We call such systems **play-out system**. The data are stored in the way in which they are transferred to the user. The data modification system has a **play-out generator** that materializes all views necessary for the play-out system.

Other applications are main-memory databases without update. The SAP database system uses a huge set of related views.

We may use the first representation for our **storage engine** and the second representation for the **input engine** or the **output engine** in data warehouse approaches.

Open problem 11.

Find techniques and theories that support treatment of redundant sets of objects and that support consistency management for sets of objects, e.g., XML document sets.

Database optimization is based on the knowledge of complexity of operations. If we know that a certain set of operations is far more complex than the rest and if we know a number of equivalent representations then we can choose among those the less complex one. A typical case of optimization is the vertical normalization where a relation is decomposed into a set of relations which has less representation complexity and which is simpler to support. Horizontal normalization selects parts of relations with lower complexity. Deductive normalization reduces relations to those elements than cannot be generated from the other elements by generation rules. At the same time we require that the representations are equivalent. So far, these three kinds of normalization are treated in a separate form.

Open problem 12.

Find a common framework for the utilization of vertical, horizontal and deductive normalization for object-relational database models.

Normalization is often based on database constraints. In order to get a correct normalization we need to know the entire set of valid constraints in the given application. This is infeasible and often not achievable.

Open problem 13.

Find a normalization theory which is robust for incomplete constraint sets.

4   Specification of Functionality

4.1   Operations for Information Systems

General operations on type systems can be defined by *structural recursion*. Given types T, T' and a collection type C^T on T (e.g. set of values of type T, bags, lists) and operations such as generalized union $\cup_{C^T}$, generalized intersection $\cap_{C^T}$, and generalized empty elements $\emptyset_{C^T}$ on C^T. Given further an element h_0 on T' and two functions defined on the types $h_1 \;\; : \;\; T \to T'$ and $h_2 \;\; : \;\; T' \times T' \to T'$.
Then we define the structural recursion by insert presentation for R^C on T as follows

$srec_{h_0,h_1,h_2}(\emptyset_{C^T}) \;=\; h_0$

$srec_{h_0,h_1,h_2}(\{\!|s|\!\}) \;=\; h_1(s)$ for singleton collections $\{\!|s|\!\}$

$srec_{h_0,h_1,h_2}(\{\!|s|\!\} \cup_{C^T} R^C) \;=\; h_2(h_1(s), srec_{h_0,h_1,h_2}(R^C))$ iff $\{\!|s|\!\} \cap_{C^T} R^C = \emptyset_{C^T}$.

All operations of the object-relational database model, the extended entity-relationship model and of other declarative database models can be defined by structural recursion, e.g.,

- selection is defined by $srec_{\emptyset,\iota_\alpha,\cup}$ for the function

$$\iota_\alpha(\{o\}) \quad = \quad \begin{cases} \{o\} \text{ if } \{o\} \models \alpha \\ \emptyset \quad \text{ otherwise} \end{cases}$$

- aggregation functions can be defined based on the two functions for null values

$$h_f^0(s) \quad = \quad \begin{cases} 0 \quad \text{ if } s = \text{NULL} \\ f(s) \text{ if } s \neq \text{NULL} \end{cases}$$

$$h_f^{\text{undef}}(s) \quad = \quad \begin{cases} \text{undef if } s = \text{NULL} \\ f(s) \quad \text{ if } s \neq \text{NULL} \end{cases}$$

through structural recursion, e.g.,

$\text{sum}_0^{null} = \quad srec_{0,h_{Id}^0,+} \;$ or $\text{sum}_{undef}^{null} = \quad srec_{0,h_{Id}^{\text{undef}},+} \;$;

$\text{count}_1^{null} = \quad srec_{0,h_1^0,+} \;$ or $\text{count}_{undef}^{null} = \quad srec_{0,h_1^{\text{undef}},+}$

or the doubtful SQL definition of the average function

$$\frac{\text{sum}_0^{null}}{\text{count}_1^{null}} \; .$$

Similarly we may define intersection, union, difference, projection, join, nesting and un-nesting, renaming, insertion, deletion, and update.

Structural recursion is also limited in expressive power. Nondeterministic while tuple-generating programs (or object generating programs) cannot be expressed.

Operations may be either used for **retrieval** of values from the database or for **state changes** within the database.

The general frame for operation definition in the co-design approach is based on views used to restrict the scope, pre-, and postconditions used to restrict the applicability and the activation of operations and the explicit description of enforced operations:

Operation φ
[View: < View_Name>]
[Precondition: < Activation_Condition >]
[Activated_Operation: < Specification >]
[Postcondition: < Acceptance_Condition >]
[Enforced_Operation: < Operation, Condition>]

The relational model and object-relational models are extended by aggregation, grouping and bounded recursion operations. The semantics of these operations varies among database management systems and has not found yet a mathematical basis [6].

Open problem 14.

Develop a general theory of extended operations for object-relational models.

4.2 Dynamic Integrity Constraints

Database dynamics is defined on the basis of transition systems. A transition system on the schema S is a pair
$$\mathcal{TS} = (\mathcal{S}, \{\xrightarrow{a}| \ a \in \mathcal{L}\})$$
where $\mathcal{S}$ is a non-empty set of state variables,
$\mathcal{L}$ is a non-empty set (of labels),
and $\xrightarrow{a} \subseteq \mathcal{S} \times (\mathcal{S} \cup \{\infty\})$ for each $a \in \mathcal{L}$.
State variables are interpreted by states. Transitions are interpreted by transactions on S.

Database lifetime is specified on the basis of paths on $\mathcal{TS}$. A *path* π through a transition system is a finite or ω length sequence of the form $s_0 \xrightarrow{a_1} s_1 \xrightarrow{a_2} ....$ The length of a path is its number of transitions.

For the transition system $\mathcal{TS}$ we can introduce now a *temporal dynamic database logic* using the quantifiers $\forall_f$ (always in the future)), $\forall_p$ (always in the past), $\exists_f$ (sometimes in the future), $\exists_p$ (sometimes in the past).

First-order predicate logic can be extended on the basis of temporal operators. The validity function I is extended by time. Assume a temporal class (R^C, l_R). The validity function I is extended by time and is defined on $S(ts, R^C, l_R)$. A formula α is valid for $I_{(R^C, l_R)}$ in ts if it is valid on the snapshot defined on ts, i.e. $I_{(R^C, l_R)}(\alpha, ts) = 1$ iff $I_{S(ts, R^C, l_R)}(\alpha, ts)$.

- For formulas without temporal prefix the extended validity function coincides with the usual validity function.
- $I(\forall_f \alpha, ts) = 1$ iff $I(\alpha, ts') = 1$ for all $ts' > ts$;
- $I(\forall_p \alpha, ts) = 1$ iff $I(\alpha, ts') = 1$ for all $ts' < ts$;
- $I(\exists_f \alpha, ts) = 1$ iff $I(\alpha, ts') = 1$ for some $ts' > ts$;
- $I(\exists_p \alpha, ts) = 1$ iff $I(\alpha, ts') = 1$ for some $ts' < ts$.

The modal operators $\forall_p$ and $\exists_p$ ($\forall_f$ and $\exists_f$ respectively) are dual operators, i.e. the two formulas $\forall_h\alpha$ and $\neg\exists_h\neg\alpha$ are equivalent. These operators can be mapped onto classical modal logic with the following definition:

$$\Box\alpha \equiv (\forall_f\alpha \wedge \forall_p\alpha \wedge \alpha);$$
$$\Diamond\alpha \equiv (\exists_f\alpha \vee \exists_p\alpha \vee \alpha).$$

In addition, temporal operators *until* and *next* can be introduced.

The most important class of dynamic integrity constraint are **state-transition con-**straints $\alpha\,O\,\beta$ which use a pre-condition α and a post-condition β for each operation O. The state-transition constraint $\alpha\,O\,\beta$ can be expressed by the the temporal formula $\alpha \xrightarrow{O} \beta$.

Each finite set of static integrity constraints can be equivalently expressed by a set of state-transition constraints $\{\wedge_{\alpha\in\Sigma}\alpha \xrightarrow{O} \wedge_{\alpha\in\Sigma}\alpha) \mid O \in Alg(M)\}$.

Integrity constraints may be enforced

- either at the procedural level by application of
    - trigger constructs [8] in the so-called active event-condition-action setting,
    - greatest consistent specialisations of operations [10],
    - or stored procedures, i.e., fully fledged programs considering all possible violations of integrity constraints,
- or at the transaction level by restricting sequences of state changes to those which do not violate integrity constraints,
- or by the DBMS on the basis of declarative specifications depending on the facilities of the DBMS,
- or a the interface level on the basis of consistent state changing operations.

Database constraints are classically mapped to transition constraints. These transition constraints are well-understood as long as they can be treated locally. Constraints can thus be supported using triggers or stored procedures. Their global interdependence is, however, an open issue.

Open problem 15.

Develop a theory of interference of database constraints that can be mapped to well-behaving sets of database triggers and stored procedures.

4.3 Specification of Workflows

A large variety of approaches to workflow specification has been proposed in the literature. We prefer formal descriptions with graphical representations and thus avoid pitfalls of methods that are entirely based on graphical specification such as the and/or traps. We use basic computation step algebra introduced in [20]:

- **Basic control commands** are sequence ; (execution of steps in sequence), parallel split $|\wedge$ (execute steps in parallel), exclusive choice $|\oplus$ (choose one execution path from many alternatives), synchronization $|^{sync}$ (synchronize two parallel threads of execution by an synchronization condition $sync$, and simple merge $+$ (merge two alternative execution paths). The exclusive choice is considered to be the default parallel operation and is denoted by $||$.

- Structural control commands are arbitrary cycles * (execute steps w/out any structural restriction on loops), arbitrary cycles $^+$ (execute steps w/out any structural restriction on loops but at least once), optional execution [] (execute the step zero times or once), implicit termination ↓ (terminate if there is nothing to be done), entry step in the step ↗ and termination step in the step ↘.

The basic computation step algebra may be extended by **advanced step commands**:

- **Advanced branching and synchronization control commands** are multiple choice $\lceil^{(m,n)}\rceil$ (choose between m and n execution paths from several alternatives), multiple merge (merge many execution paths without synchronizing), discriminator (merge many execution paths without synchronizing, execute the subsequent steps only once) n-out-of-m join (merge many execution paths, perform partial synchronization and execute subsequent step only once), and synchronizing join (merge many execution paths, synchronize if many paths are taken, simple merge if only one execution path is taken).
- We also may define **control commands on multiple objects** (CMO) such as CMO with a priori known design time knowledge (generate many instances of one step when a number of instances is known at the design time), CMO with a priori known runtime knowledge (generate many instances of one step when a number of instances can be determined at some point during the runtime (as in FOR loops)), CMO with no a priori runtime knowledge (generate many instances of one step when a number of instances cannot be determined (as in a while loop)), and CMO requiring synchronization (synchronization edges) (generate many instances of one activity and synchronize afterwards).
- **State-based control commands** are deferred choice (execute one of the two alternative threads, the choice which tread is to be executed should be implicit), interleaved parallel executing (execute two activities in random order, but not in parallel), and milestone (enable an activity until a milestone has been reached).
- Finally, **cancellation control commands** are used, e.g. cancel step (cancel (disable) an enabled step) and cancel case (cancel (disable) the case).

These control composition operators are generalizations of workflow patterns and follow approaches developed for Petri net algebras.

Operations defined on the basis of this general frame can be directly translated to database programs. So far no theory of database behavior has been developed that can be used to explain the entire behavior and that explain the behavior in depth for a run of the database system. A starting point for the development of such theory might be the proposal [17] to use abstract state machines [2].

Open problem 16.

Develop a theory of database behavior. This theory explains the run of database management systems as well as the run of database systems.

4.4   Architecture of Database Engines

Operating of information systems is modelled by separating the systems state into four state spaces:

$\mathcal{E}R^C$ = (input states $\mathcal{IN}$, output states $\mathcal{OUT}$,
engine states $\mathcal{DBMS}$, database states $\mathcal{DB}$).

The input states accommodate the input to the database system, i.e. queries and data. The output space allow to model the output of the DBMS, i.e. output data of the engine and error messages. The internal state space of the engine is represented by engine states. The database content of the database system is represented in the database states. The four state spaces can be structured. This structuring is reflected in all four state spaces. For instance, if the database states are structured by a database schema then the input states are accordingly structured.

Using value-based or object-relational models the database states can be represented by relations. An update imposed to a type of the schema is in this case a change to one of the relations. State changes are modelled on the basis of abstract state machines [2] through *state change rules*. An engine is specified by its programs and its control. We follow this approach and distinguish between

programs that specify units of work or services and meet service quality obligations and

control and coordination that is specified on the level of program blocks with or without *atomicity and consistency* requirements or specified through *job control commands*.

Programs are called with instantiated parameters for their variables. Variables are either static or stack or explicit or implicit variables. We may use furthermore call parameters such as `onSubmit` and `presentationMode`, priority parameters such as `onFocus` and `emphasisMode`, control parameters such as `onRecovery` and `hookOnProcess`, error parameter such as `onError` and `notifyMode`, and finally general transfer parameters such as `onReceive` and `validUntil`.

Atomicity and consistency requirements are supported by the variety of transaction models. Typical examples are flat transactions, sagas, join-and-split transactions, contracts or long running activities [16].

State changes $T(s_1, ..., s_n) := t$ of a sub-type T' of the database engine $\mathcal{E}R^C$. A set $\mathcal{U} = \{T_i(s_{i,1}, ..., s_{i,n_i}) := o_i | 1 \leq i \leq m\}$ of object-based state changes is *consistent*, if the equality $o_i = o_j$ is implied by $T_i(s_{i,1}, ..., s_{i,n_i}) = T_j(s_{j,1}, ..., s_{j,n_j})$ for $1 \leq i < j \leq m$.

The *result* of an execution of a consistent set $\mathcal{U}$ of state changes leads to a new state $\mathcal{E}R^C$ to $\mathcal{E}R^C + \mathcal{U}$

$$(\mathcal{E}R^C + \mathcal{U})(o) = \begin{cases} Update(T_i, s_{i,1}, ..., s_{i,n_i}, o_i) \\ \quad \text{if } T_i(s_{i,1}, ..., s_{i,n_i}) := o_i \in \mathcal{U} \\ \mathcal{E}R^C(o) \\ \quad \text{in the other case} \end{cases}$$

for objects o of $\mathcal{E}R^C$.

A parameterized programm $r(x_1, ..., x_n) = P$ of arity n consists of a program name r, a transition rule P and a set $\{x_1, ..., x_n\}$ of free variables of P.

An information system $\mathcal{E}R^C$ is a **model** of ϕ ($\mathcal{E}R^C \models \phi$) if $[\![\phi]\!]_\zeta^{\mathcal{E}R^C} = true$ for all variable assignments ζ for free variables of ϕ.

Two typical program constructors are the execution of a program for all values that satisfy a certain restriction

— FOR ALL x WITH ϕ →⬚ DO P ⬚→

and the repetition of a program step in a loop

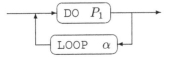

We introduce also other program constructors such as sequential execution, branch, parallel execution, execution after value assignment, execution after choosing an arbitrary value, skip, modification of a information system state, and call of a subprogram.

We use the abstract state machine approach also for definition of semantics of the programs.

A transition rule $\mathcal{P}$ leads to a set $\mathcal{U}$ of state changing operations in a state $\mathcal{E}R^C$ if it is consistent. The state of the information system is changed for a variable assignment ζ to $\mathtt{yields}(\mathcal{P}, \mathcal{E}R^C, \zeta, \mathcal{U})$.

Semantics of transition rules is defined in a calculus that uses rules of the form

$$\frac{\text{prerequisite}_1, ..., \text{prerequisite}_n}{\text{conclusion}} \quad \text{where condition}$$

For instance, the state change imposed by the first program step is defined by

$$\forall a \in I : \text{yields}(\mathcal{P}, \mathcal{E}R^C, \zeta[x \mapsto a], \mathcal{U}_a)$$
$$\text{yields}(\text{FOR ALL } x \text{ WITH } \phi \text{ DO } \mathcal{P}, \mathcal{E}R^C, \zeta, \bigcup_{a \in I} \mathcal{U}_a)$$

$$\text{where} \quad I = range(x, \phi, \mathcal{E}R^C, \zeta)$$

The range $range(x, \phi, \mathcal{E}R^C, \zeta))$ is defined by the set $\{o \in \mathcal{E}R^C | [\![\phi]\!]^{\mathcal{E}R^C}_{\zeta[x \mapsto a]} = true\}$.

Open problem 17.

Develop a general theory of abstraction and refinement for database systems that supports architectures and separation or parqueting of database systems into components.

4.5 Data Extraction Frameworks

Surprises, data warehousing, and complex applications often require sophisticated data analysis. The most common approach to data analysis is to use data mining software or reasoning systems based on artificial intelligence. These applications allow to analyse data based on the data on hand. At the same time data are often observational or sequenced data, noisy data, null-valued data, incomplete data, of wrong granularity, of wrong precision, of inappropriate type or coding, etc. Therefore, brute-force application of analysis algorithms leads to wrong results, to loss of semantics, to misunderstandings etc.

We thus need general frameworks for data analysis beyond the framework used for data mining. We may use approaches known from mathematics for the development of data analysis framework.

Open problem   18.

Develop semantic-preserving and pragmatic-preserving frameworks for data extraction.

5   Specification of Interactivity

Interactivity of information systems has been mainly considered on the level of presentation systems by the Arch or Seeheim separation between the application system and the presentation system. Structuring and functionality are specified within a database modelling language and its corresponding algebra. Pragmatics is usually not considered within the database model. The interaction with the application system is based on a set of views which are defined on the database structure and are supported by some functionality.

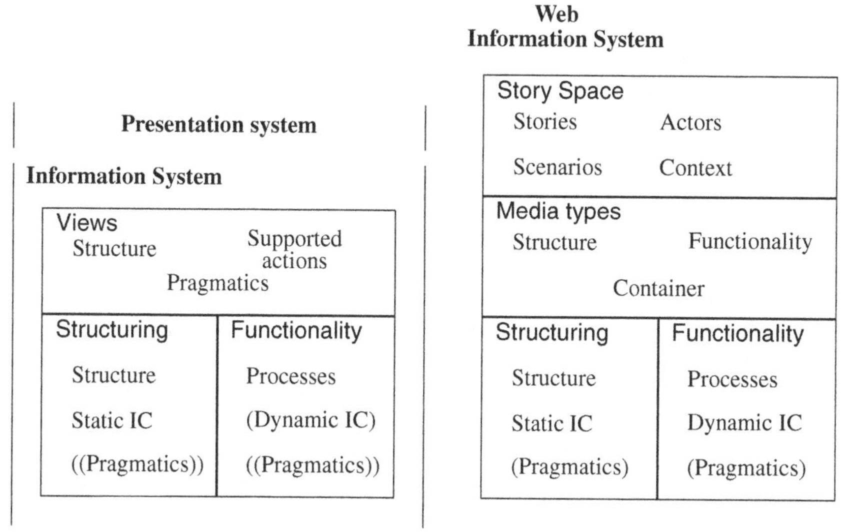

Web Information System

Presentation system		Story Space	
		Stories	Actors
Information System		Scenarios	Context
Views	Supported	Media types	
Structure	actions	Structure	Functionality
	Pragmatics		Container
Structuring	Functionality	Structuring	Functionality
Structure	Processes	Structure	Processes
Static IC	(Dynamic IC)	Static IC	Dynamic IC
((Pragmatics))	((Pragmatics))	(Pragmatics)	(Pragmatics)

Fig. 3. Architecture of Information Systems Enhanced by Presentation Systems Compared With the Architecture of Web Information Systems

The general architecture of a web information system is shown in Figure 3. This architecture has successfully been applied in more than 30 projects resulting in huge or very large information-intensive websites and in more than 100 projects aiming in building large information systems.

In the co-design framework we generalize this approach by

introduction of media objects which are generalized views that have been extended
    by functionality necessary, are adapted to the users needs and delivered to the actor
    by a container [12] and by

introduction of story spaces [13] which specify the stories of usage by groups of
users (called actors) in their context, can be specialized to the actual scenario of
usage and use a variety of play-out facilities.

User interaction modelling involves several partners (grouped according to characteristics; group representatives are called 'actors'), manifests itself in diverse activities
and creates an interplay between these activities. Interaction modelling includes modelling of environments, tasks and actors beside modelling of interaction flow, interaction
content and interaction form.

5.1 Story Space

Modelling of interaction must support multiple scenarios. In this case, user profiles,
user portfolios, and the user environment must be taken into consideration. The *story*
of interaction is the intrigue or plot of a narrative work or an account of events. The
language SiteLang [20] offers concepts and notation for specification of story spaces,
scene and scenarios in them.

Within a story one can distinguish threads of activity, so-called *scenarios*, i.e., paths
of scenes that are connected by transitions. We define the story space Σ_W as the 7-tuple
$(S_W, T_W, E_W, G_W, A_W, \lambda_W, \kappa_W)$ where S_W, T_W, E_W, G_W and A_W are the set of
scenes created by W, the set of scene transitions and events that can occur, the set
of guards and the set of actions that are relevant for W, respectively. Thus, T_W is a
subset of $S_W \times S_W$. Furthermore $\lambda_W : S_W \rightarrow SceneSpec$ is a function associating
a scene specification with each scene in S_W, and $\kappa_W : T_W \rightarrow E_W \times G_W \times A_W$,
$t \mapsto (e, g, a)$ is a function associating with each scene transition t occurring in W
the event e that triggers transition t, the guard g, i.e. a logical condition blocking the
transition if it evaluates to false on occurrence of e, and the action a that is performed
while the transition takes place.

We consider scenes as the conceptual locations at which the interaction, i.e., dialogue takes place. Dialogues can be specified using so-called dialogue-step expressions.
Scenes can be distinguished from each other by means of their identifier: Scene-ID.
With each scene there is associated a media object and the set of actors that are involved in it. Furthermore, with each scene a representation specification is associated
as well as a context. Scenes therefore can be specified using the following frame:

Scene = (Scene-ID
    DialogueStepExpression
    Data views with associated functions
    User
        UserID
        UserRight
        UserTasksAssigned
        UserRoles
    Representation (styles, defaults, emphasis, ...)
    Context (equipment, channel, particular)

Dialogue-step expressions consist of dialogues and operators applied to them. A typical scene is displayed in Figure 4. A learner may submit solutions in the data mining

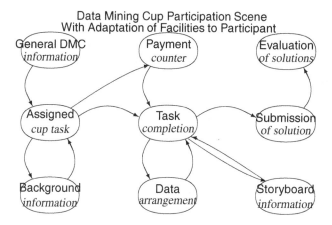

Fig. 4. One of the Scenes for Active Learning

cup. Before doing so, the user must pay a certain fee if she has not already been paying. The system either knows the user and his/her profile. If the user has already paid the fee then the payment dialogue step is not shown. If the user has not paid the fee or is an anonymous user then the fee dialogue step must be visited and the dialogue step for task completion is achievable only after payment.

5.2   Media Type Suite

Media types have been introduced in [12]. Since users have very different needs in data depending on their work history, their portfolio, their profile and their environment we send the data packed into containers. Containers have the full functionality of the view suite. Media type suites are based on view suites and use a special delivery and extraction facility. The media type suite is managed by a system consisting of three components:

Media object extraction system: Media objects are extracted and purged from database, information or knowledge base systems and summarized and compiled into media objects. Media objects have a structuring and a functionality which allows to use these in a variety of ways depending on the current task.

Media object storage and retrieval system: Media objects can be generated on the fly whenever we need the content or can be stored in the storage and retrieval subsystem. Since their generation is usually complex and a variety of versions must be kept, we store these media objects in the subsystem.

Media object delivery system: Media objects are used in a large variety of tasks, by a large variety of users in various social and organizational contexts and further in various environments. We use a media object delivery system for delivering data to the user in form the user has requested. Containers contain and manage the set of media object that are delivered to one user. The user receives the user-adapted container and may use this container as the desktop database.

This understanding closely follows the data warehouse paradigm. It is also based on the classical model-view-control paradigm. We generalize this paradigm to media objects, which may be viewed in a large variety of ways and which can be generated and controlled by generators.

Open problem 19.

Provide a theory that support adaptable (to the user, to the context, to the history, to the environment) delivery of content and reception of content.

6   Integrating Specification Aspects into Co-design

The languages introduced so far seem to be rather complex and the consistent development of all aspects of information systems seems to be rather difficult. We developed a number of methodologies to development in order to overcome difficulties in consistent and complete development. Most of them are based on top-down or refinement approaches that separate aspects of concern into abstraction layers and that use extension, detailisation, restructuring as refinement operations.

6.1   The Abstraction Layer Model for Information Systems Development

We observe that information systems are specified at different abstraction layers:

1. The *strategic layer* addresses the purpose of the information system, i.e. its mission statement and the anticipated customer types including their goals. The results of the design process are conducted into the *stakeholder contract specification*.
2. The *requirements elaboration layer* describes the information system, analyzes business processes and aims in elicitation of the requirements to the information system. The results of the design process are combined into the *system specification*.
3. The *business layer* deals with modelling the anticipated usage of the information system in terms of customer types, locations of the information space, transitions between them, and dialogues and discourses between categories of users (called actors). The result of this abstraction layer is compiled into an *extended system manual* including mockups of the interfaces and scenarios of utilization.
4. The *conceptual layer* integrates the conceptual specification of structuring, functionality, distribution and interactivity. The results of this step are the database schema, the workflows, the view and media type suites, the specification of distribution, and the story space.
5. At the *implementation layer*, logical and physical database structures, integrity enforcement procedures, programs, and interfaces are specified within the language framework of the intended platform. The result of specification at the implementation layer is the *implementation model*. This model is influenced by the builder of the information system.
6. The exploitation layer is not considered here. Maintenance, education, introduction and administration are usually out of the scope of conceptualization of an application.

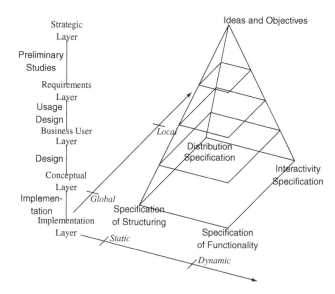

Fig. 5. The Abstraction Layer Model of the Database Design Process

6.2   The Co-design Methodology

Methodologies must confirm both with the SPICE v. 2.0 and SW-CMM v. 2.0 requirements for consistent system development. The co-design framework is based on a step-wise refinement along the abstraction layers. Since the four aspects of information systems - structuring, functionality, distribution and interactivity - are interrelated they cannot be developed separately. The methodology sketched below is based on steps in the following specification frame.

The steps used in one of the methodologies are:

Strategic layer
1. Developing visions, aims and goals
2. Analysis of challenges and competitors

Requirements elaboration layer
3. Separation into system components
4. Sketching the story space
5. Sketching the view suite
6. Specifying business processes

Business user layer
7. Development of scenarios of the story space
8. Elicitation of main data types and their associations
9. Development of kernel integrity constraints, e.g., identification constraints
10. Specification of user actions, usability requirements, and sketching media types
11. Elicitation of ubiquity and security requirements

Rule # i **Name of the step**	Task 1. Task 2. ...
Used documents	Documents of previous steps (IS development documents) Customer documents and information
Documents under change	IS development documents Contracts
Aims, purpose and subject	General aims of the step Agreed goals for this step Purpose Matter, artifact
Actors involved	Actor A, e.g., customer representatives Actor B, e.g. developer
Theoretical foundations	Database theory Organization theory Computer science Cognition, psychology, pedagogic
Methods and heuristics	Syntax and pragmatics Used specification languages Simplification approaches
Developed documents Results	IS development documents Results and deliverables
Enabling condition for step	Information gathering conditions fulfilled by the customer side Information dependence conditions Conditions on the participation
Termination condition for step	Completeness and correctness criteria Sign-offs, contracts Quality criteria Obligation for the step fulfilled

Conceptual layer

    12. Specification of the story space

    13. Development of data types, integrity constraint, their enforcement

    14. Specification of the view suite and distribution

    15. Development of workflows

    16. Control of results by sample data, sample processes, and sample scenarios

    17. Specification of the media type suite

    18. Modular refinement of types, views, operations, services, and scenes

    19. Normalization of structures

    20. Integration of components along architecture

Implementation layer

    21. Transformation of conceptual schemata into logical schemata, programs, and interfaces

    22. Development of distribution

    23. Developing solutions for performance improvement, tuning

24. Transformation of logical schemata into physical schemata
25. Checking durability, robustness, scalability, and extensibility

The co-design methodology has been practically applied in a large number of information system projects and has nevertheless a sound theoretical basis. We do not want to compete with UML but support system development at a sound basis without ambiguity, ellipses and conceptual mismatches.

Open problem 20.

Develop quality characteristics and measurements for each of the modelling steps and measures for quality preserving transformations.

7 Conclusion

Database and information systems research has brought up a technology that is now a part of the everyday infrastructure. Database systems are running as embedded systems, e.g. for car information systems, as collaborating systems or as stand-alone systems. The technology has got a maturity that allows to use and to install a database system whenever information-intensive computation supports an application. Due to this wide range of applications information systems are currently extended to support distributed computation and web information systems. These novel directions of research are currently in intensive change and attract a lot of research. Within this paper we developed a proposal for extending classical database technology to these novel areas. This proposal is only one way of ending and applying the current technology.

At the same time a number of problems remains still open in the area of databases. We have summarized the achievements of a theory of database structuring and functionality. Some of these achievements become already neglected with the development of novel information system models. They are, however, neatly combinable with the novel models. In the first part of the paper we summarized these achievements and introduced some of the main open problems of current database research.

References

1. Beeri, C., Thalheim, B.: Identification as a primitive of database models. In: Polle, T., Ripke, T., Schewe, K.-D. (eds.) FoMLaDO 1998. Proc. Fundamentals of Information Systems, 7th Int. Workshop on Foundations of Models and Languages for Data and Objects, Timmel, Ostfriesland, pp. 19–36. Kluwer, London (1999)
2. Börger, E., Stärk, R.: Abstract state machines - A method for high-level system design and analysis. Springer, Berlin (2003)
3. Demetrovics, J., Molnar, A., Thalheim, B.: Graphical and spreadsheet reasoning for sets of functional dependencies. In: Atzeni, P., Chu, W., Lu, H., Zhou, S., Ling, T.-W. (eds.) ER 2004. LNCS, vol. 3288, pp. 54–66. Springer, Heidelberg (2004)
4. Jaakkola, H., Thalheim, B.: Software quality and life cycles. In: Eder, J., Haav, H.-M., Kalja, A., Penjam, J. (eds.) ADBIS 2005. LNCS, vol. 3631, pp. 208–220. Springer, Heidelberg (2005)

5. Kaschek, R.: Konzeptionelle Modellierung. PhD thesis, University Klagenfurt, Habilitations-schrift (2003)

6. Lenz, H.-J., Thalheim, B.: OLAP databases and aggregation functions. In: 13th SSDBM 2001, pp. 91–100 (2001)

7. Lenz, H.-J., Thalheim, B.: OLTP-OLAP schemes for sound applications. In: Draheim, D., Weber, G. (eds.) TEAA 2005. LNCS, vol. 3888, pp. 99–113. Springer, Heidelberg (2006)

8. Levene, M., Loizou, G.: A guided tour of relational databases and beyond. Springer, Berlin (1999)

9. Lockemann, P.: Information system architectures: From art to science. In: Proc. BTW 2003, pp. 1–27. Springer, Berlin (2003)

10. Schewe, K.-D.: The specification of data-intensive application systems. PhD thesis, Brandenburg University of Technology at Cottbus, Faculty of Mathematics, Natural Sciences and Computer Science, Advanced PhD Thesis (1994)

11. Schewe, K.-D., Thalheim, B.: Fundamental concepts of object oriented databases. Acta Cybernetica 11(4), 49–81 (1993)

12. Schewe, K.-D., Thalheim, B.: Modeling interaction and media objects. In: Bouzeghoub, M., Kedad, Z., Métais, E. (eds.) NLDB 2000. LNCS, vol. 1959, pp. 313–324. Springer, Heidelberg (2001)

13. Srinivasa, S.: An algebra of fixpoints for characterizing interactive behavior of information systems. PhD thesis, BTU Cottbus, Computer Science Institute, Cottbus (April 2000)

14. Suppes, P.: Representation and invariance of scientific structures. CSLI publications, Stanford (2002)

15. Thalheim, B.: Open problems in relational database theory. Bull. EATCS 32, 336–337 (1987)

16. Thalheim, B.: Entity-relationship modeling – Foundations of database technology. Springer, Berlin (2000)

17. Thalheim, B.: ASM specification of internet information services. In: Moreno-Díaz Jr., R., Buchberger, B., Freire, J.-L. (eds.) EUROCAST 2001. LNCS, vol. 2178, pp. 301–304. Springer, Heidelberg (2001)

18. Thalheim, B.: Co-design of structuring, funtionality, distribution, and interactivity of large information systems. Computer Science Reports 15/03, Cottbus University of Technology, Computer Science Institute (2003)

19. Thalheim, B.: The conceptual framework to user-oriented content management. In: EJC 2006, Trojanovice (May 2006)

20. Thalheim, B., Düsterhöft, A.: Sitelang: Conceptual modeling of internet sites. In: Kunii, H.S., Jajodia, S., Sølvberg, A. (eds.) ER 2001. LNCS, vol. 2224, pp. 179–192. Springer, Heidelberg (2001)

21. Vinogradov, I.: Mathematical encyclopaedia (in 5 volumes). Soviet Encyclopaedia, Moscov (in Russian) (1982)

Remark: Our main aim has been the survey of current database research. We restrict thus the bibliography only to those references which are necessary for this paper. An extensive bibliography on relevant literature in this field can be found in [16].

Metaphor Modeling on the Semantic Web

Bogdan D. Czejdo[1], Jonathan Biguenet[2], and John Biguenet[3]

[1] Center for AMEDD Strategic Studies
Fort Sam Houston, TX 78234-5047
bczejdo@hotmail.com
[2] Tulane University
New Orleans, LA 70118
jbiguene@tulane.edu
[3] Loyola University
New Orleans, LA 70118
biguenet@loyno.edu

Abstract. Metaphor is a high-level abstract concept that can be an important part of active conceptual modeling. In this paper, we use the extended Unified Modeling Language (UML) for metaphor modeling. We discuss how to create UML diagrams to capture knowledge about metaphors. The metaphor-based processing system on the Semantic Web can support new query/search operations. Such a computer system can be used for a broad spectrum of applications such as predicting surprises (e.g., terrorist attacks) or generating automatically new innovations.

1 Introduction

Recently a new framework for active conceptual modeling was introduced [20]. This framework can be very useful for performing "situation awareness and monitoring for military forces, business enterprises" [20] and other organizations [20]. Such a new challenge requires not only proper integration of existing techniques but also introduction of new methodologies and tools. In this paper we propose to use metaphor-based computer systems for active conceptual modeling. The conceptual metaphor itself has served civilization very well for many centuries. It created better situation awareness for humans and triggered many innovations. It was one of the very important indicators of human intelligent behavior. Unfortunately, it was also sometimes related to negative behavior. To summarize, metaphorical thinking has led to many "surprises," both positive and negative. If we could identify, model and document these surprises by a computer system, we might be able to make significant in-roads into artificial intelligence and generate automatically new innovations and predict negative surprises.

Currently most of the data in an explicit or hidden form is available on the Web. Very intensive practical and theoretical research is still required to uncover this hidden information and create a Semantic Web [1]. The Semantic Web research includes not only how to extract valuable information from natural language texts but also how to create a new method for storing information that will allow for easy access to all concepts and the relationships between them. For the latter, one of the

P.P. Chen and L.Y. Wong (Eds.): ACM-L 2006, LNCS 4512, pp. 97–111, 2007.
© Springer-Verlag Berlin Heidelberg 2007

problems is that different types of highly abstract concepts may require new modeling constructs. The metaphor is one of them [21].

In this paper, we describe metaphor modeling and its implications for query/search systems. We have found that creating a proper model for higher-level abstract concepts such as metaphor requires many new constructs beyond those described in previous papers [2, 3, 4, 5, 6, 7, 8, 16, 17]. In addition, building these constructs requires many different types of interactions between the computer and a human.

Here, we use notation similar to that of Universal Modeling Language (UML) [2, 12] for metaphor modeling. The UML, though created and used for software design, is not specific to software design. This kind of modeling has been already used for a long time to represent a variety of object-relationship systems [13]. UML modeling is characterized by concise graphical notation, offers availability of simple tools, and can be easily extended to use informal specifications. It can be very useful for ontology specifications, especially in the initial phase when all details are not clear. Therefore, UML diagrams can still play an important role even though new notations such as OWL [15] and new tools such as Protégé [18] and SWOOP [19] allow for powerful operations on ontologies including queries not readily available in typical UML tools.

Once the metaphor-based processing systems are built, then the Semantic Web can be used for new diverse applications such as predicting surprises (e.g., terrorist attacks) or generate automatically new innovations. Many other applications can be significantly improved such as assistance in English-literature education.

2   Metaphor and Simile

Metaphor plays a major role in literature, and poetry in particular. For example, in "Letters from a Poet Who Sleeps in a Chair," Nicanor Parra, the Chilean poet, offers a self-evident metaphor: "The automobile is a wheelchair" ("*El automóvil es una silla de ruedas*"). Though a reader will immediately grasp its irony, what would a computer capable of processing natural language make of this metaphor?

A number of issues must be resolved before that question can be answered. Like many a high-school poetry student, the computer might be stymied by the difference between a metaphor and a simile. The use of *like* or *as* turns a metaphor into a simile. So "the automobile is like a wheelchair" is a simile and, perhaps, easier for a student or a computer to grasp than its statement as a metaphor. In the simile, the automobile is merely an approximation of a wheelchair, sharing a limited number of attributes. "Like" signals simply a congruence between the two entities compared; its use eliminates the possible confusion of the metaphor: that the two entities share identity.

But the metaphor, while demanding a similar cognitive operation, declines to limit explicitly the relationship between the two entities compared to mere approximation. Instead, the metaphor hints at shared identity. In mathematical terms, we are tempted to express the metaphor as A equals B. The difficulty the computer faces in decoding such a statement's intended meaning is obvious. The poet does not mean to say that an automobile equals a wheelchair with the same unreserved equality of A and B in the above formula. Rather, the metaphor demands an ironic interpretation, in which the reader recognizes a presupposed context for the asserted relationship between the automobile and the wheelchair. The humor of the metaphor—and its force—anticipate such active interpretation by the reader. Can a computer be taught to get the joke of Parra's metaphor?

3 Requirements for a Tool to Model Metaphor

For a diagram to work well for modeling metaphor, it needs to satisfy several requirements. First, it needs to allow the capture of preliminary information about classes (or objects), their properties, and relationships between them. Second, it should allow for systematic refinement of the diagram to clarify the meanings, remove the redundancy, and resolve contradictions. Third, it should be expandable in the sense that new high-level operators would be allowed to be defined.

Let us discuss the first requirement for diagrams for modeling metaphors: the need to allow the user to capture preliminary information in the initial phase of knowledge specification when all details typically are not yet clear. The need of this phase can be compared to top-down natural language processing by humans when various relationships are established between natural language sentences or sentence fragments. These sentences or sentence fragments initially might have the ambiguous or unclear meanings, redundancy, and contradictions within them. [14].

Let us introduce a UML-like diagram that satisfies this property. The diagram consists of boxes corresponding to objects or classes. Each box has three areas. The first area is an object/class name, the second area contains a list of properties written as sentence fragments in English, and the third area contains the list of functions written again as sentence fragments in English as shown in Fig. 1. Since such diagrams are very close to UML diagrams, for simplicity we will designate them as such.

The UML diagram shown in Fig. 1. is used as a running example in our paper. This diagram describes the Vehicle and how it relates to a Human. We define a vehicle as an object that is operated by a human, that is used by the operator to move, and that expands the possible movements of an operator. The relationship to Human is also presented as an association to Human and is represented by the sentence "A human operates a vehicle." We define Human here (in a limited way) as an object that has a body and moves. Both Vehicle and Human are broken down into subclasses. For Vehicle, we have the two categories, Automobile and Wheelchair. An Automobile we define as an object that has four wheels, has seats, is powered by an engine, carries multiple passengers, is primarily used by a non-handicapped human, and, as a type of Vehicle, moves its operator (a non-handicapped human) around. A Wheelchair, on the other hand, is described in this diagram as an object which has four wheels, a seat, is powered by a human, carries a single passenger, is used by a handicapped human, and, also as a vehicle, moves its operator around. On the topic of the Vehicle operator, Human has two subtypes. A Non-Handicapped Human is one that moves within a large space. A Handicapped Human is one in whom a component or components of the body do not function and who moves in a limited space.

The second requirement for the metaphor modeling diagram is to allow the user refinement of the diagram to clarify the meanings, remove the redundancy, and resolve contradictions. Our UML diagram satisfies this requirement. An extensive discussion on how to use our UML diagram for knowledge refinement is discussed in [3]. The third requirement for the metaphor modeling diagram will be discussed in the next section when we expand our UML diagram to include new high-level relationships such as metaphor relationship.

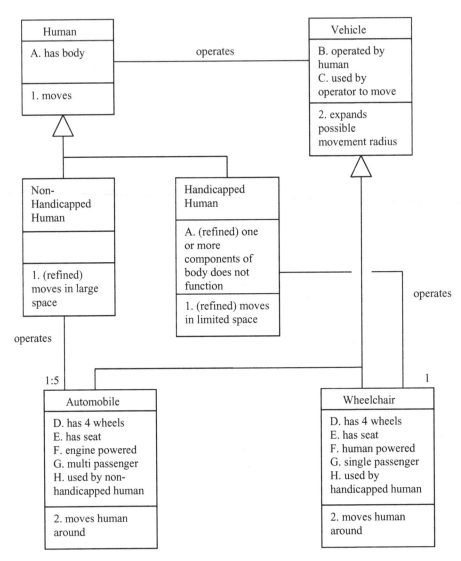

Fig. 1. Common Knowledge about Vehicles Represented by a UML-like Class Diagram

The fourth purpose of the second requirement for the metaphor modeling diagram is the possibility to use the diagram to answer both simple queries and complex queries. The underlying UML diagram once constructed contains the knowledge about a specific subject or many subjects (e.g., about Vehicles). Therefore, it can be used to answer simple queries in these subject areas such as "Describe Vehicle." The result will include all properties and functions of the class (object). Among complex queries, the comparison queries are some of the most useful for this project.

The comparison queries should allow for an extended comparison of classes. Comparisons of classes would require comparison of their properties and functions as well as of all relationships. An example of a comparison query is "Compare an automobile with a wheelchair." As a result of the comparison, all common properties and functions would be identified. In addition, the common relationships need to be identified. In our case, we have a common relationship to the concept "wheeled vehicle." Since the relationship type is a super-class, some additional processing is necessary. All attributes from a super-class can be inherited, so they are also indicated as common. In addition, any super-class of the initial super-class needs to be identified and all properties and functions marked as common.

4 Metaphor Modeling by UML Diagram

The metaphor in a natural language can have a very complex meaning not only stressing the similarity between two classes (or objects) but also emphasizing or de-emphasizing some properties, functions, and relationships. Let us model the metaphor described in section 2, "The automobile is a wheelchair" ("*El automóvil es una silla de ruedas*"), by a UML diagram.

We noted that the metaphor hints at shared identity and symmetrical structure. A literary interpretation tells us that is not really the case. So we need to define metaphor more precisely. In terms of UML modeling, the metaphor is rather a non-symmetrical special subclass/superclass relationship that allows for modifying properties, functions, and relationships of one concept by another. We will call this special subclass/superclass relationship a "metaphor relationship" and indicate it by double lines on the diagram. Let us consider the metaphor from our example. We can create a new relationship type called "metaphor" between Automobile and Wheelchair in the direction from the Automobile to Wheelchair as shown in Fig. 1.

The defining of a metaphor in the UML diagram means not only insertion of a new type of relationship called "metaphor" but also identification of the list of modified properties, functions, and relationships with respect to the given metaphor, as well as determination of the type of modification of each listed property, function, and relationship. The modification can be to emphasize, de-emphasize, or superimpose. The first two are self-explanatory, but the superimposition requires some discussion. Superimposing does not erase the real values but only temporarily masks them with new values. This masking process takes place only when we consider metaphor relationships. We refer to such properties, functions, and relationships as "superimposed" (in our example in Fig. 2, we will mark the property "multi-passenger" by "superimposed" and store the new value "single-passenger" to indicate its new meaning for the metaphor).

Finally, we must also identify any propagating metaphors (i.e., metaphors created on the basis of another metaphor). In our example, we have a propagating metaphor between human and handicapped human derived from our initial metaphor between an automobile and a wheelchair. Fig. 2 also reflects that specification. Once the derived metaphor is stated, all of the same steps discussed above must be taken to define it.

Let us consider queries for metaphors. When we request the definition of Automobile, the system will respond with a few statements, and one of these will be "Automobile is Wheelchair." In addition, we can enrich the queries by including the

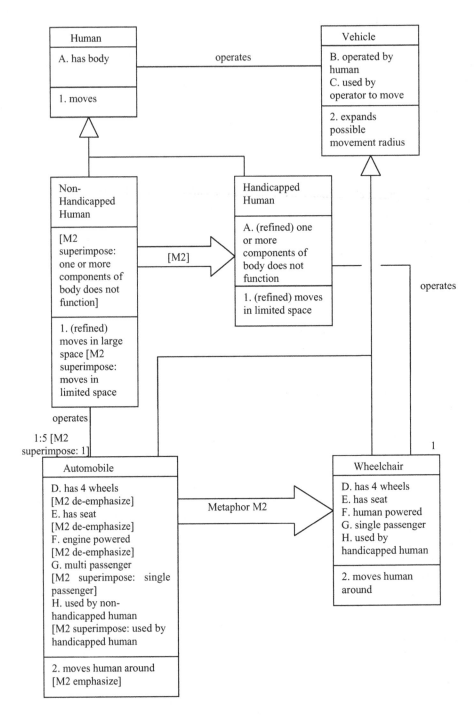

Fig. 2. Metaphor about Vehicles Represented by a UML Class Diagram

question "What are the attributes with changed value for Automobile?" The answer could be "multi-passenger," but the metaphor argues that it is also "single-passenger."

5   A Metaphor-Based System for the Semantic Web

A system for processing of ontologies with metaphor can be created as shown in Fig. 3. The users of the system can be a metaphor analyst and a metaphor query/search user. The metaphor analyst has access to both ontology (e.g., represented by our UML diagram) and a Semantic Web, whose number are growing dramatically each year. Based on the comparison operator, the metaphors can be defined and stored within the local ontology. At the same time, the relationships between concepts in our ontology and Semantic Web are established in the form of pointers or indexes. Based on this information, a metaphor query/search user can issue requests for ontology and/or Semantic Web. If requests are for the Semantic Web, then they are processed in two phases. First the relevant information is retrieved from our ontology, and then the links to Semantic Web are identified and the appropriate information retrieved. Results of requests can be transparent for the user in the sense that users do not need to be aware if the answer is provided based on ontology or involves the Semantic Web search.

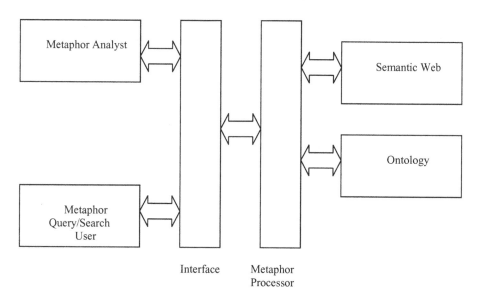

Fig. 3. The Architecture of Software System for Processing of Ontologies with Metaphor Several subsystems can participate in the metaphor processing (Metaphor Processor)

5.1   A Simple Sub-system for Defining Metaphors

A simple software sub-system for processing of metaphors is based on maximum user involvement in the definition of metaphor. This is the most straightforward method of defining a metaphor, and it involves the following phases:

1. Stating the metaphor. This can be done by a simple interface shown in Fig. 4 for the "automobile is a wheelchair" metaphor. It is equivalent to the creation of new high-level relationship between Automobile and Wheelchair.

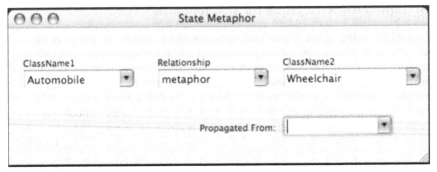

Fig. 4. The Knowledge Database Interface to Declare the Metaphor

2. Identifying similarity of attribute types. This step can be done by a simple interface shown in Fig. 5 for a different metaphor: "A computer is a file cabinet." This phase is equivalent to re-labeling of some of the attributes, if necessary, to show their similarity (e.g., attribute B and F are the same types of attributes that define storage media in general). The values of the attributes are different, but that will be handled in the next step.

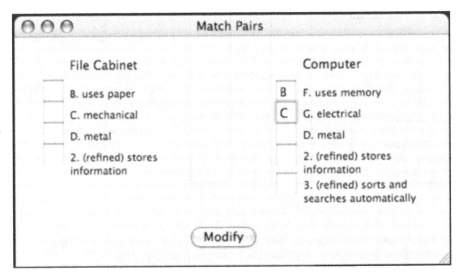

Fig. 5. The Knowledge Database Interface to Match the Attributes for the Metaphor

3. Defining the role of matching attributes for the metaphor. This can be done by a simple interface shown in Fig. 6 for the "automobile is a wheelchair" metaphor. This phase is equivalent to tagging the attributes with a role ("Superimpose," "Emphasize," or "De-Emphasize"). In this example, Attributes D-F are tagged with "De-Emphasize."

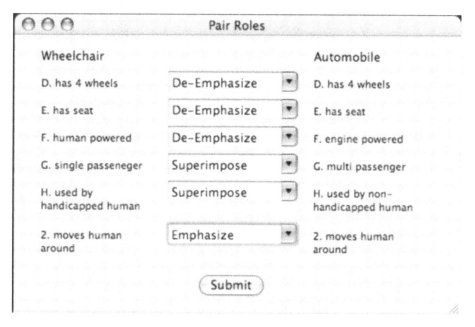

Fig. 6. The Knowledge Database Interface to Find the Roles for the Attributes in the Metaphor

4. Explaining the non-matching attribute role for the metaphor. This can be done by another simple interface. There is only one non-matching attribute role for the metaphor: "De-Emphasize." In a case when the attribute cannot be ignored, there is a need to redefine the model to find the matching attribute.

5.2 A Computer-Assisted Sub-system for Defining Metaphors

There are many types of assistance that can be provided to the metaphor analyst. The system can be used as a verification tool. In this mode, the system would provide verification of the analyst steps (e.g., identifying the problem of attributes dangling in objects with no matched pair as shown in Fig. 7). It is inevitable that sometimes there will be an object that has an attribute type that does not appear in the other object. For example, an attribute dealing with type of motion is essential to the description of a vehicle but totally unnecessary in a description of a table. In general, it must be decided whether these left-over properties and functions should be de-emphasized. If not, it indicates a need to redefine the metaphor because otherwise the information carried in the relationship between this dangling attribute and the other object (which is obviously too important to ignore) will weaken the metaphor.

The system can be used as a guidance tool. Instead of using interfaces in the sequence selected by the user, the analyst can be guided in the process. In this mode, the system would ask specific questions of the analyst to organize the creation and analysis of a metaphor and lead the student step-by-step through the process. For

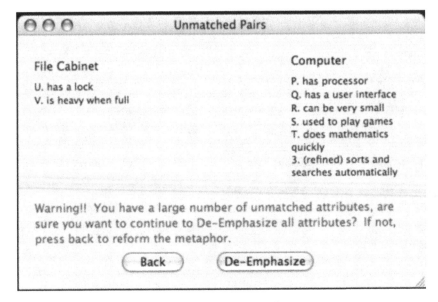

Fig. 7. The Ontology Interface

example, after defining high-level relationships between Automobile and Wheelchair, the analyst can be asked questions about possible propagation of the metaphor to between Non-handicapped Human and Handicapped Human.

5.3 An Automatic Sub-system for Defining Metaphors

Good preparation is the key for automatic metaphor detection, especially good preparation of attributes including cleaning and integration. Ideally, we would like come up with the scheme for global integration of all attributes (i.e., identify identical attributes, identify similar attributes, and define the type of and level of similarity). Then we would like to do the same with relationships. However, as we know from much less complex information systems, attribute, relationship, and object integration is still a challenging problem in many cases (e.g., coding scheme used for different attributes that can not be easily converted).

There are two types of automatic metaphor discovery. The first method is based on metaphor discovery by algorithmic analysis (comparison operator) of objects and their relationships. The success of this method depends heavily on the precision of describing attributes, the compatibility of their descriptions, using the same coding system, etc. It can work well in well-defined and integrated ontologies.

Since many ontologies do not satisfy these strict requirements, metaphor discovery by quantitative analysis very often needs to be employed. This method will also be called metaphor mining. In this method, first there are some measures of how good metaphors are developed. The measures are based on the number of matching attributes, including the number of the attributes that are de-emphasised and the number of attributes that need to be superimposed. The measures are also based on the number of matching relationships, including the number of relationships that need to be de-emphasised, the number of relationships that need to be superimposed, and recursively computed propagation measures. A version of the metaphor mining system could learn from good examples of metaphors and use the extracted features to identify other metaphors.

5.4 A Metaphor Query/Search Sub-system

Once metaphors are recorded in our ontology, we can use the ontology for query/search operations. Generally there are two types of query/search sub-systems: a system returning all metaphors and a system returning only the most relevant metaphors. The first type of system would be used in well-developed and integrated ontologies, and the results would be relatively precise (as precise as the identification of the metaphor was). The second type of system would be used for not-well-integrated ontologies, and the results would contain ranked metaphors. That type of system usually has the possibility of false positives (high-ranked metaphors that are not really good metaphors) and true negatives (good metaphors that are not listed at all or with very low rank).

Theoretically the method for defining the metaphor is not related to the type of query/search system, but practically it can be very important. Let us elaborate on that topic. A simple sub-system, a computer-assisted sub-system or an automatic sub-system based on algorithmic analysis for defining metaphors, does not need to assign ranking for metaphor. It means that without any enhancement, a metaphor query/search sub-system will return all requested metaphors.

An automatic sub-system based on metaphor mining by definition assigns ranking for candidate metaphors in order to decide about the final set of metaphors based on some threshold. It means that a metaphor-based query/search sub-system can have this ranking available and use it to return requested metaphors with highest rankings. The threshold can be a very important parameter of a query/search. The query request with a high threshold would mean that only the "strongest" metaphors according to some measure would be returned. Generally a threshold can be multidimensional, each dimension corresponding to a different perspective for the metaphor.

The implementation of the metaphor query/search system poses another interesting question. Especially for small ontologies, the metaphors can be identified beforehand and the query/search system would simply retrieve them. For example, if all metaphors for "Flying Bomb" are identified, including a passenger airplane with a full tank of gas, then the query can be specified in some language to retrieve these metaphors. The language can be designed based on SQL. Let us call such a language MetaphorSQL. Its simplex syntax contains three clauses: (1) SELECT clause to choose the object of interest or its attributes, (2) FROM clause to define all objects or classes of objects participating in the query and being involved in some type of relationship (a higher-level relationship as is metaphor or a simple relationship based on join operations), and (3) WHERE clause to specify restriction on any object or class of objects. Let us consider as an example a query to retrieve all objects that can reach the Twin Towers and that can act as (be a metaphor of) a "Flying Bomb." Such a query can be specified in MetaphorSQL as:

```
SELECT X.name
FROM X is_metaphor Flying_Bomb
WHERE X can_reach "Twin_Towers"
```

For large ontologies and the Semantic Web, the preprocessing might not be realistic for all cases. Then the metaphors need to be identified "on fly" based on a very narrow search condition such as X can_reach "Twin_Towers." In such cases, a ranking clause (4) RANK should also be added to make the search more selective.

6 Applications for Metaphor-Based Systems

Since metaphor is a very powerful concept used extensively by humans, it is expected that metaphor-based computer systems will play a crucial role in the further development of computer tools for our civilization. Metaphors are highly compressed packets of information. They are used to speak about topics for which we have no well-defined words. This implies that a system that analyzes metaphors is, in a sense, an automatic decoding tool. For instance, messages sent via the Internet could be scanned to reveal the meanings of purposely vague wordings. Further, changes made to web content could be used to generate possible meanings: what is the relationship between the old word/phrase and the new one? Even if the two fragments are fairly ambiguous, it is possible that the relationship between them is not.

6.1 Application of Metaphor-Based Systems in Education

Metaphor-based systems can be very helpful for students and teachers of any subject that requires heavy use of metaphors. The system can show teachers the possible difficulties the human students may face in learning new material involving high-level abstract concepts [9, 10, 11]. In many disciplines, such as literature, those difficulties are often obscured by traditional teaching methods. For example, English teachers frequently depend upon classroom discussion to help students develop skill in reading poetry. Though students may be able to offer a general analysis of the poem that can be critiqued by the instructor, few exercises reveal the process by which they have come to an understanding (or a misunderstanding) about even a basic poetic element such as a metaphor.

By modeling metaphors, we have been able to observe the steps by which a computer tries to make sense of the complex meaning of a simple metaphor. The difficulties of that enterprise parallel the problems a student faces in trying to achieve understanding. We have reached a number of conclusions about the ways in which a student processes a comparison, such as a literary metaphor. This approach may lead to a broader application in which the modeling of metaphor reveals how a teacher-constructed ontology will serve to identify subject-specific learning problems teachers may need to address in the classroom.

The system can also be used as a guidance or verification tool. In the guidance mode, the system would ask specific questions of the student to organize the creation and analysis of a metaphor and lead the student step-by-step through the process.

Once the metaphor has been added to the ontology, it contains the relevant knowledge about the subject. Therefore it can be used to answer queries and provide explanations about the metaphor. Another simple interface can be used to accomplish that task. Such an interface can provide a systematic method to query students about the defined metaphor.

Depending on the level of the students, the Web-based learning can be adapted to focus discussion on local ontology or more advanced aspects of Web ontologies. For example, if the students are just learning the concept of metaphor, it is enough for the system to go through the steps of creating the metaphor using the prepared local ontology. If, on the other hand, the students already understand the basics, the students might spend more time integrating various Web ontologies for the purpose of metaphor specification and metaphor propagation.

6.2 Application of Metaphor-Based Systems in Surprise Detection

The English-literature example also teaches us about the critical role of metaphor for surprising the reader. Let us refer to written metaphors as passive metaphors helping us understand better the properties of objects and their real or hidden properties. Let us refer to active metaphors of an object as a new usage of the object beyond its primary function or goal. There are many such surprises—such as terrorist attacks, military surprises, etc.

Surprises are most of the time active metaphors. Therefore, surprise understanding and detection can be significantly supported by metaphor-based systems. We can create a "knowledge base" of all active metaphors (i.e., surprises in history including military surprises). Based on such a "knowledge base," the new surprises can be predicted better. For example, terrorist attacks can be predicted when they introduce new uses of everyday objects (tools) as carriers of weapons of destruction.

7 Summary

In this paper, we have discussed metaphor as a very important high-level abstract concept that can lead to a new generation of metaphor-based processing systems. Such systems can be used in a variety of situations (e.g., to predict surprises or to teach literature, and poetry in particular, to list just two). For metaphor modeling, we used UML-like diagrams that gave us more flexibility than other notations typical for a Semantic Web, such as OWL. Once the ontologies with metaphors are readily available on a given subject, the usability of a future Semantic Web will be significantly extended.

There are many difficult problems related to the changes of metaphor or changes of the underling ontology that need to be addressed to achieve a full capability of dynamic systems. For example, how important is the change of metaphor and how can it be captured? Also, can metaphor support the "delta" model that represents the difference between two snapshots of the ontology?

Note

The views expressed in this article are those of the authors and do not reflect the views of the Army Medical Department, Department of the Army, or Department of Defense.

References

1. Bemers-Lee, T., Hendler, J., Lassila, 0.: The Semantic Web. Sci. Am. 284(5), 34–43 (2001)
2. Booch, G., Rumbaugh, J., Jacobson, I.: The Unified Modeling Language User Guide. Addison Wesley, Reading, MA (1999)
3. Czejdo, B., Mappus, R., Messa, K.: The Impact of UML Diagrams on Knowledge Modeling, Discovery and Presentations. Journal of Information Technology Impact 3(1), 25–38 (2003)
4. Czejdo, B., Czejdo, J., Lehman, J., Messa, K.: Graphical Queries for XML Documents and Their Applications for Accessing Knowledge in UML Diagrams. In: Proceedings of the First Symposium on Databases, Data Warehousing and Knowledge Discovery, Baden Baden, Germany, pp. 69–85 (2003)
5. Czejdo, B., Czejdo, J., Eick, C., Messa, K., Vernace, M.: Rules in UML Class Diagrams. Opportunities of Change. In: Proceedings of the Third International Economic Congress, Sopot, Poland, pp. 289–298 (2003)
6. Delteil, A., Faron, C.: A Graph-Based Knowledge Representation Language. In: Proceedings of the 15th European Conference on Artificial Intelligence (ECAI), Lyon, France (2002)
7. Dori, D., Reinhartz-Berger, I., Sturm, A.: OPCAT—A Bimodal CASE Tool for Object-Process Based System Development. In: ICEIS 2003. Proceedings of the IEEE/ACM 5th International Conference on Enterprise Information Systems, Angers, France, pp. 286–291 (2003), http://www.ObjectProcess.org
8. Lehrnan, F. (ed.): Semantic Networks in Artificial Intelligence. Pergamon, Oxford, UK (1999)
9. Mayer, R.E.: Multimedia Learning. Cambridge University Press, New York (2002)
10. McTear, M.F. (ed.): Understanding Cognitive Science. Ellis Horwood, Chichester, UK (1998)
11. Novak, J.D., Gowin, D.B.: Learning How to Learn. Cambridge University Press, New York (1984)
12. Rumbaugh, J., Jacobson, I., Booch, G.: The Unified Modeling Language Reference Manual. Addison-Wesley, Reading, MA (1999)
13. Rumbaugh, J.: Objects in the Constitution: Enterprise Modeling. Journal of Object Oriented Programming (1993)
14. Schulte, R., Biguenet, J.: Theories of Translation. The University of Chicago Press, Chicago (1992)
15. Smith, M.K., McGuinness, D., Volz, R., Welty, C.: Web Ontology Language (OWL) Guide Version 1.0. W3CWorking Draft (2002)
16. Sowa, J.F.: Conceptual Structures: Information Processing in Mind and Machine. Addison-Wesley, Reading, MA (1984)
17. Sowa, I.F.: Knowledge Representation: Logical, Philosophical, and Computational Foundations. Brooks Cole, Pacific Grove, CA (1999)

18. Gennari, M.A., Musen, R.W., Fergerson, W.E., Grosso, M., Crubezy, H., Eriksson, N.F., Noy, S.W.T.: The Evolution of Protégé: An Environment for Knowledge-Based Systems Development (2002)
19. Kalyanpur, A., Parsia, B., Sirin, E., Cuenca-Grau, B., Hendler, J.: Swoop - a web ontology editing browser, Journal of Web Semantics 4 (1) (2005)
20. Chen, P.P., Wong, L.Y.: A Proposed Preliminary Framework for Conceptual Modeling of Learning from Surprises. In: Arabnia, H.R., Joshua, R. (eds.) ICAI 2005. Proceedings of the 2005 International Conference on Artificial Intelligence, Las Vegas, Nevada, USA, June 27-30, 2005, vol. 2, pp. 905–910. CSREA Press (2005)
21. Thalheim, B., Dusterhoft, A.: The Use of Metaphorical Structures for Internet Sites. Data and Knowledge Engineering Journal 35(2), 161–180 (2000)

Schema Changes and Historical Information in Conceptual Models in Support of Adaptive Systems

Luqi[1] and Douglas S. Lange[2]

[1] Naval Postgraduate School, Department of Computer Science
Monterey, CA
http://www.nps.navy.mil/cs/facultypages/luqi/home.html
[2] Space and Naval Warfare Systems Center
53560 Hull Street
San Diego, CA 92152, USA
luqi@nps.edu, doug.lange@navy.mil

Abstract. Conceptual changes and historical information have not been emphasized in traditional approaches to conceptual modeling such as the entity-relationship approach. Effective representations for such changes are needed to support robust machine learning and computer-aided organizational learning. However, these aspects have been modeled and studied in other contexts, such as software maintenance, version control, software transformations, etc. This paper reviews some relevant previous results, shows how they have been used to simplify conceptual models to help people make sense out of complex changing situations, and suggests some connections to conceptual models of machine learning. Areas where research is required to support conceptual models for adaptive systems are also explored. These are suggested by studies of the issues surrounding deployment of adaptive systems in mission critical environments.

1 Introduction

The motivating context for studying active conceptual learning is to provide a systematic framework for learning from surprises that can support machine learning. Envisioned complex applications include adaptive command and control, situation monitoring for homeland security, and assessment of preparedness for coping with disasters. These complex applications require learning mechanisms that are robust in the sense of being able to accommodate qualitative unplanned changes in the view of the world, which go beyond adjusting parameters in a predetermined model or structure. Also of concern is organizational learning, which involves collective and collaborative learning by a group of humans and intelligent software agents. Among the new issues in this context are communication in support of learning, representation of communal knowledge, and analysis of past information in future contexts relative to recorded past events to derive improved policies and decisions for current action and future planning. Various conceptual models and knowledge representations enable the application of machine learning techniques. This paper explores past work on modeling of information changes and explores its applicability in this context.

The entity-relationship (E-R) model is one of the oldest and most successful of the approaches for computer modeling of conceptual information for describing the real

P.P. Chen and L.Y. Wong (Eds.): ACM-L 2006, LNCS 4512, pp. 112–121, 2007.
© Springer-Verlag Berlin Heidelberg 2007

world. The same basic structures have been used for database design as well as the design of complex systems. For example, many current approaches to object-oriented design, and popular notations for this purpose, such as UML, use many of the concepts and constructs from the E-R approach for conceptual modeling of object oriented software designs.

Many of the knowledge representations used to support expert systems and machine learning also have aspects similar to the E-R model. Among these are semantic networks, frames, and scripts. Entities and relationships play a prominent part in all these knowledge representations.

In these approaches to knowledge representation, the most obvious difference from the E-R model is the explicit and sometimes sophisticated use of inheritance. Some knowledge representation approaches tailor the inheritance mechanism to account for exceptional cases in which a specialization may not quite have all of the characteristics of the inherited general pattern. Such situations are relatively common in real complex situations, where few general rules are truly universal. A standard example is that flight is a distinguishing characteristic of birds; however, there exist flightless birds such as penguins, kiwis and ostriches. This kind of inheritance differs somewhat from the mathematically simpler kinds of specialization commonly used in database and software modeling, due to the above mentioned possibility of exceptional cases. It is also common for knowledge representations to support default or tentative values for attributes. Such values are plausible but may not be entirely certain, and can be overridden when contradictory information with stronger supporting evidence becomes available. Uncertainty of information and possible exceptions to empirical rules are relevant for machine learning, particularly if we are concerned with real-world situation monitoring and learning from surprises.

Dynamic behavior, state changes, and changes in conceptual structures over time are typical characteristics of the scenarios that appear in the motivating context. These aspects have not been extensively studied in the context of the E-R model, which focuses on data representation issues and does not include a data manipulation model, as do many of the other conceptual models used in database design and intelligent systems. These aspects of the application domain have been studied in other contexts. We examine some of the insights and approaches to modeling changes that have come out of those other contexts, and explore possible correspondences with extensions of the E-R approach suited to active conceptual learning.

Some of the relevant contexts include object-oriented modeling, version control, software evolution, software change merging, and software reuse.

2 Dynamics in UML

Modeling state changes is an integral part of software design, and has been accommodated in the popular notations used for this purpose, such as UML. The kinds of state changes emphasized by these notations do not quite match the needs of active conceptual learning because they focus mainly on state changes that occur within a given conceptual model, rather than changes between different models. In other words, the changes that have been most commonly studied and modeled occur mainly on the instance level rather than at the scheme level.

For example, UML uses state diagrams to explicitly represent state changes, emphasizing the finite state aspects of control systems, particularly for those that have a combination of discrete components (typically software or digital hardware) and continuous components (typically analog electronics or physical systems). It also uses sequence diagrams to partially and implicitly describe subsystem state changes implicit in patterns of behavior, which may not be expressible using finite state models.

The usual interpretations of these constructs do not match the needs of active conceptual learning very well because they all fall within the scope of a single conceptual scheme. Even in the more sophisticated applications of state machine modeling, where subsystem interfaces can have different type signatures and qualitatively different behaviors in different states, the kinds of state changes that can be naturally represented do not span the range needed to effectively support machine learning.

The kind of state-based modeling supported by UML and its relatives is suited for modeling different modes of operation in real-time systems. Some of these modes may correspond to degraded levels of service due to partial hardware failures in robust systems designed so that the most critical functions will continue to be provided even if some parts of the system cease functioning. Other plausible applications for this kind of state model include different modes that correspond to different missions or different external circumstances. In the latter case, the benefit is that different real-time schedules can be used under different conditions to make more effective use of limited computing resources during peak loads, especially if the peak loads have different characteristics in different modes of operation. In both cases, the purpose of the modes is really to make the system predictable, in the sense of guaranteeing certain levels of reliability or performance, while allowing additional flexibility under the actual operating conditions that can be realized by mode changes. Each mode represents and anticipated and pre-analyzed situation. The predictability is attained by static analysis of a finite set of states that are fixed and known at design time.

In summary, there is a mismatch between the common UML constructs and our context because active conceptual learning seeks to adapt to new situations that were not explicitly anticipated and the UML constructs were not intended to do this. The rest of this paper examines other models of change that were developed to accommodate less predictable situations.

3 Version Control

One source of models for unpredictable changes in complex environments is version control for software systems. The environment of a typical system is a set of people, social organizations, and physical phenomena. Such environments are typically complex, and are subject to changes that are not accurately predictable or practically boundable. The information models developed for this domain have some common characteristics:

- Archival orientation. New versions can be added, but existing versions are frozen and in principle have unbounded lifetimes. This is appropriate for conceptual schemas, because we never want to lose the ability to understand and manipulate our historical data.

- Discrete time. Each version in the repository represents a fixed snapshot of the situation at a given point in time. The points in time are defined by the completion of new versions, and are typically irregularly spaced.
- Transitions. Some models include representations of the development activities that produce each version [1, 2]. These activities are represented as special derivation dependency relationships among versions, which typically have attributes. Examples of such attributes appropriate for the context of system development include who developed the new version and how many person-hours were spent. For a disaster relief context, transitions would represent disasters, and relevant attributes might include the number of casualties and a set of issues that must be addressed by the disaster relief team. A dependency relation of this kind records which versions of an object were derived from which other versions of the same or different objects. These dependency relations are strict partial orderings that are consistent with the temporal ordering. If A depends on B then A was created after B was created. The creation event for a version marks the end point of the development activity that produced the new version and corresponds to a unique point in time.
- Branching history. Version models typically can represent parallel lines of system evolution that can be interpreted as different configurations of a system or different products in a product line. A line of development is a chain (a totally ordered subset of versions) with respect to the derivation dependency ordering introduced in the previous bullet. In the context of learning, parallel branches could represent alternative courses of action related to a situation or tentative hypotheses about an uncertain situation along with associated inferences and plans.
- Hierarchical structure. Versions typically have subcomponents, corresponding to the decomposition structure of a system design. This structure is a special kind of aggregation relationship with associated coherency constraints. For example, every input to a subsystem must either be an input to the parent system or an output from a sibling subsystem. In the context of learning, a modeled concept could have sub-concepts that are needed for its explanation or that provide supporting evidence for a modeled belief.

Version models consist of disciplined kinds of entities and relationships, and are consistent specializations of E-R models with additional properties.

This kind of model has an analog to scheme changes, which is used to model changes in system structure in different versions. Engineers naturally strive to accommodate small changes to system requirements by changing the properties of a single component, without affecting the overall architecture of the system. Thus the structure of a version is usually the same as the structure of the previous version. However, occasionally requirements change in ways that require fundamental changes to the architecture, involving new interfaces and new paths of communication within the system. Such a structural change corresponds to a scheme change in a database. Similar patterns may be useful for representing internal "paradigm shifts" resulting from major new "insights" gained by a machine learning system. Some of the mathematical properties of such changes, and principles for combining such changes can be found in [3, 4].

The addition of requirements tracing or managing multiple deployed configurations adds further examples of how software engineering knowledge management tools can provide lessons for ACM-L. Models link the various pieces of information to the phases of the software lifecycle and allow traceability and change detection based on dependencies. Examples of these dependencies are the links between criticisms of the software to the issues that are raised. The issues can be linked to requirements, which can be linked to specification changes, and ultimately to implementation and changes in implementation. Test plans and results can be linked in, as can other artifacts such as training materials. Previous research [8] has shown that not only can tools be developed that track such information, but decision trees can be generated from the information that can provide insight into the relative complexities of decisions that must be made in the software evolution process.

The models of software engineering can take the form of a hypergraph [2]. A possible abstract form of such a model is shown in the following diagram.

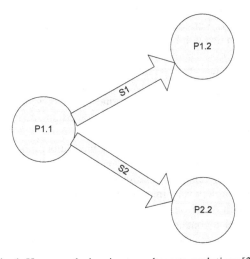

Fig. 1. Hypergraph showing two alternate evolutions [2]

Expanding one of the steps from the figure above, shows us the involvement of multiple items of information in the evolution step, as shown in the figure below.

The decomposition edges d1.1 and d2.2 show that A1.1 and B1.1 are parts of P1.1, and that A2.2, B1.1, and C2.1 are parts of P2.2. The substep s2.a2 derives A2.2 from A1.1, while the substep s2.c1 derives the new component C2.1 from nothing at all [2].

One possible schematic model of such processes could be as shown in the following diagram. Considering the information and the processes in this light shows how a dynamic information model can be created in a software evolution process.

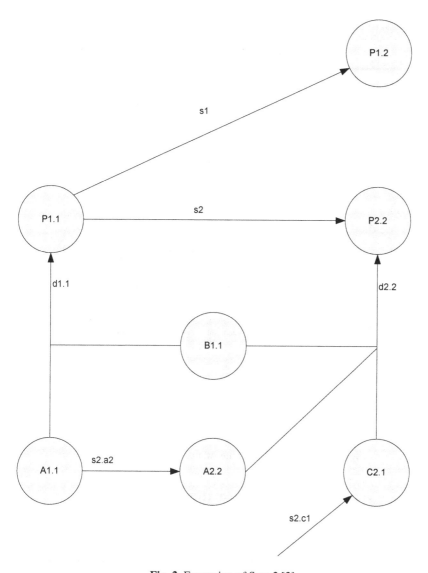

Fig. 2. Expansion of Step 2 [2]

Similar models can be created for product line management, where specific requirements are tied to the configuration code and delivered products where variations
must be created to support multiple customers or installations [10].

One of the issues when combining knowledge fragments, or changes to a knowledge base, is that information derived from different sources can conflict. Some of the
models for combining changes include representations for different kinds of conflicts,
rules for locating conflicts, and in some cases even rules for resolving conflicts. When
active conceptual learning is used to support teams of human analysts, this type of
knowledge modeling may have relevance.

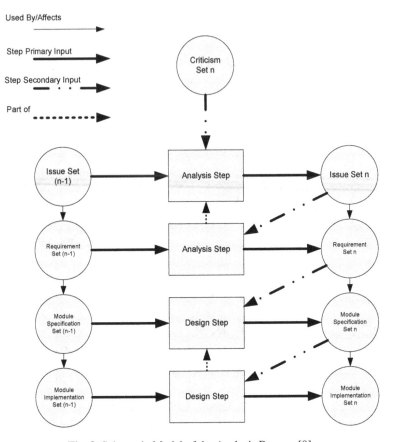

Fig. 3. Schematic Model of the Analysis Process [9]

4   Software Evolution and Transformations

One of the problems in software evolution is to make sense out of changes. One approach to providing assistance for this process is to explain a system's development as if it were developed by a perfect process, without the false starts and dead ends that characterize the development of real software systems.

One approach to doing this that may have some applicability to active conceptual learning is a mathematical model that enables us to formally "rewrite history" as suggested above [5]. The basic idea rests on a relationship between different versions of a software artifact that represents an upwards compatible change (hopefully an improvement over the previous version!). Upwards compatible means that the new version meets the requirements of the previous version, both syntactically and semantically, so that the new version can be substituted for the old version in any well-formed context without doing any harm. An example of how such relationships can be concretely defined and used to rewrite history as described above can be found in

[5]. The advantage of such rewriting is to compute a simplified view of history that includes only the parts relevant to explaining the issue at hand.

In the context of modeling data there can be several different notions of upwards compatibility. A simple idea is to consider the semantics of a schema to consist of the set of all possible instances. In such a case, a change from version *v1* of a schema to version *v2* would be considered upwards compatible if every possible instance of *v1* is also an instance of *v2*. Some examples of upwards compatible changes to a schema include adding new entity types or new relationships. Examples of changes that would not be upwards compatible are deleting or renaming types or changing a binary relationship into a ternary relationship. Under this point of view a change *c1* that adds a new type and change *c2* that adds a new relationship among previously existing types would be both upwards compatible and compatible with each other. If change *c1* added a new type and change *c2* added another new type with different properties but the same name, we would have an example of a conflict between two changes.

There are many subleties involved if we wish to have a more robust view of compatiblity that can bridge substantial variations in how the same reality is modeled. A very simple example is motivated by changes that add new attributes. If we take the above formulation literally, such a change would not be upwards compatible. However, if we introduce some standards regarding null values and allow some natural transformations, this can become upwards compatible change. The transformation needed treats an instance without an arbitrary attribute as if it were equivalent to one that has a null value for that attribute. This enables a natural transformation from instances of the old schema, none of which have values for the new attribute, to corresponding instances of the new schema, which have null values for the new attribute. Note that similar transformations could make type renaming upward compatible.

We conjecture that similar techniques may be able to highlight the subset of a complex situation that is relevant to a particular purpose. This may be one way that automated processes can help humans find some of the needles in a world full of haystacks.

5 Software Reuse

The intended applications for active conceptual learning include situation monitoring, which includes recognizing patterns in monitored behavior. Prior work on software component search for the purpose of reuse had to address some similar phenomena, and suggests some problems that should be addressed when applying E-R models to active conceptual learning [6, 7].

Software components have complex behaviors that are difficult to describe. Experience shows that different people tend to describe the same system behavior in different ways. This applies not only to informal communication, but to formal modeling as well: even at an abstract level, the structure, type signatures and axiomatic descriptions of different interfaces for realizing the same behavior can be radically different in different formalizations. Bridging these differences brings difficulties: theoretically the desired equivalences are not in general decidable, and practically, matching procedures must be constrained to terminate within relatively short time windows, resulting in poor ability to recognize an appreciable fraction of the matches that are actually present in the software knowledge base. We found a somewhat counter-intuitive result: methods based on matching concrete behaviors for a small number of test cases performed much better in terms of precision and recall (standard quality of service metrics for information retrieval) than theoretically better founded methods based on general pattern matching and inference. The latter methods had better precision (rejection of false positives), but this advantage was far outweighed by their poor recall (high incidence of false negatives).

We suggest that instance-level matching may have similar advantages over scheme-level matching in the context of finding patterns in situations.

6 Conclusions

The evolution of software (as generalized through all of the topics discussed in previous sections) is in some ways like the evolution of any knowledge. Active conceptual modeling will need to provide many of the same features that models supporting software evolution require, and the mathematical constructs developed for software evolution can inform active conceptual modeling research. Tracking and controlling schema changes, recognizing different knowledge from different time periods, combining knowledge from different sources or experiences, utilizing updated knowledge to reevaluate or re-experience past situations are all among the uses of an active conceptual model. Similarly, the constructs that bind software modifications to the processes that created them is similar to the need for knowledge to be tied to the events from which that knowledge was derived.

However, there are areas where the software engineering research discussed above does not satisfy the needs for active conceptual models. The large grain discrete time model for software development needs to be evaluated for suitability for this new purpose. Relations in software engineering are usually considered to be fully certain, but uncertainty must be dealt with in active conceptual models. These and other extensions can positively inform software engineering models as well.

Active conceptual learning can benefit from representations extending the E-R approach. Software engineering research has helped to demonstrate that by using E-R based models in automating aspects of the evolution of software. Areas where extensions to E-R models are needed include formalization of changes, the rules governing how changes combine, and methods for detecting when situation changes should induce corresponding scheme changes. Uncertainty and time models must be incorporated as well.

References

1. Luqi: A Graph Model for Software Evolution. IEEE Transactions on Software Engineering 16(8), 917–927 (1990)
2. Luqi, Goguen, J.: Formal Methods: Promises and Problems. IEEE Software 14(1), 73–85 (1997)
3. Dampier, D., Luqi, Berzins, V.: Automated Merging of Software Prototypes. Journal of Systems Integration 4(1), 33–49 (1994)

4. Berzins, V., Dampier, D.: Software Merge: Combining Changes to Decompositions. Journal of Systems Integration (Special Issue on Computer Aided Prototyping) 6(1-2), 135–150 (1996)
5. Berzins, V., Luqi, Yehudai, A.: Using Transformations in Specification-Based Prototyping. IEEE Transactions on Software Engineering 19(5), 436–452 (1993)
6. Goguen, J., Nguyen, D., Messeguer, J., Luqi, Zhang, D., Berzins, V.: Software Component Search. Journal of Systems Integration (Special Issue on Computer Aided Prototyping) 6(1-2), 93–134 (1996)
7. Steigerwald, R., Luqi, McDowell, J.: A CASE Tool for Reusable Software Component Storage and Retrieval in Rapid Prototypng. Information and Software Technology 38(9), 698–706 (1991)
8. Berzins, V., Ibrahim, O., Luqi: A Requirements Evolution Model for Computer Aided Prototyping. In: Proceedings of the 9th International Conference on Software Engineering and Knowledge Engineering, Madrid, Spain, pp. 38–47 (1997)
9. Ibrahim, O.: A Model and Decision Support Mechanism for Software Requriements Engineering, doctoral dissertation, Naval Postgraduate School (1996)
10. McGregor, J.: The Evolution of Product Line Assets, Technical Report, CMU/SEI-2003-TR-005, Carnegie-Melon University, Software Engineering Institute (2003)

Using Active Modeling in Counterterrorism

Yi-Jen Su, Hewijin C. Jiau, and Shang-Rong Tsai

Electrical Engineering Department
National Cheng Kung University, Tainan, Taiwan 70101
iansu@ee.ncku.edu.tw, {jiauhjc,srtsai}@mail.ncku.edu.tw

Abstract. Terrorist organizations attain their goals by attacking various targets to jeopardize human lives and intimidate governments. As new terrorist attacks almost always aim to *break the mold* of old plots, tracing the dynamic behaviors of terrorists becomes crucial to national defense. This paper proposes using active modeling in analyzing unconventional attacks in the design of counterterrorism system. The intelligent terrorism detection system not only detects potential threats by monitoring terrorist networks with identified threat patterns, but also continually integrates new threat features in terrorist behaviors and the varying relationships among terrorists.

Keywords: Terrorist Network, Active Modeling.

1 Introduction

Since the September 11th terrorist attack happened in 2001, how to apply information technologies to detect potential threats from the behavior of suspicious terrorists becomes one of the most important issues for homeland security. The behavior of terrorists and the relationship between them are always highly dynamic and hard to predict. To avoid the detection of intelligence organizations, terrorist groups change their connection structure from traditional hierarchies to networks. Most of them are forming loosely coupled cells [1], like the Leninist cell and the Maoist cell.

Generally, there are some limitations in analyzing the relationship between terrorist networks [2]. First, the information about all suspicious terrorists may be *incomplete*. Even after a thorough investigation, the terrorist analyst might still fail to obtain the whole picture of a certain terrorist group's activities. Furthermore, the *fuzzy boundary* problem causes difficulties when the analyst wants to decide whom to include and whom not to include in a terrorist cell. The terrorist network is also *dynamic* because new members might be recruited and some members might be captured.

In [3], conventional conceptual modeling focuses primarily on *static* relationships between entities. It can not effectively model the changes of terrorists' behaviors as well as the dynamic and time-varying relationships among them. Active modeling, a new conceptual modeling approach with a continually learning process, not only describes all aspects of a problem domain, but seeks to adapt the conceptual model from different perspectives [12]. Clearly active modeling can satisfy the dynamic requirements of counterterrorism. On the one hand, it models past incidents as

P.P. Chen and L.Y. Wong (Eds.): ACM-L 2006, LNCS 4512, pp. 122–131, 2007.
© Springer-Verlag Berlin Heidelberg 2007

scenarios to predict the attacks under plotting; on the other, it adapts the current model to detect future attacks by incorporating new clues that have been ignored in historical incidents or appear in new kinds of attacks.

2 Background

The main purpose of terrorists is to produce fear in humans by launching surprise attacks. When terrorists start to plan an attack, they identify potential targets, plan methods of attack, and calculate the response such an attack will garner from their intended audience. To counteract, terrorism analysts adopt "connecting the dots" to get the whole picture of incidents when an attack happens. The approach is like a snowball sampling method. It recursively includes all related intelligences to understand the cause and effect of an event.

How to effectively model the structure of terrorist organizations is the key to understanding the behavior of terrorists. Terrorist groups will not stop developing new kinds of attacks and improving the efficiency of existing methods. Therefore, only modeling the static relationship between terrorists is not enough. We still have to face the challenges of dynamic relationships between terrorists.

2.1 Terrorist Network

The terrorist network forms distributed cohesive cells. All cell members remain in close contact with each other to provide emotional support and to prevent desertion or breach of security procedures. The cell leader as a gatekeeper is the only person who communicates and coordinates with higher levels and other cells.

For example in [2], Krebs used a social network to model the relationships between the 19 hijackers in the 911 attack event. He manually mapped the terrorists' interactions found in public newspapers into a terrorist network that had an average path length of 4.75 steps between any two terrorists. He also successfully identified Mohamed Atta as the ring leader in this attack by using the centrality metrics (Degrees, Closeness, and Betweenness) of Social Network Analysis (SNA).

Basically, a terrorist network, as a heterogeneous social network, has various kinds of relations between its members [6]. One or more kinds of relations exist between pairs of terrorists, representing interaction, kinship, ownership, and trust [5]. It can be modeled as an Attributed Relational Graph (ARG) [11] which includes a set of actors that may have relationships with one another. In general, the actors in an ARG stand for people, organizations, stuff, or events.

In order to launch attacks that generate maximum fear and avoid be detected, all terrorist attacks are plotted to *break the mold* [5]. Even if we can smoothly model terrorists' activities into ARGs and identify frequent substructures within terrorist attacks that have already happened by Graph-based Data Mining (GDM) [8][9], these discovered patterns might still fail to help us detect potential terrorist attacks. In this system, therefore, *inexact graph matching* [5] will be applied to detect potential threats that approximate the discovered patterns in suspicious terrorist groups (*cells*).

2.2 Active Modeling

The major difference between conventional conceptual modeling and active modeling is that the latter provides closer conceptualization of reality for global understanding and communication by learning [3][12]. When the collected data of the problem domain are incomplete or when similar events are left out from the record of historical events, the omitted events will result in surprises. While static conceptual modeling can not represent unconventional events well, the active modeling adopts continually learning mechanism to solve this problem.

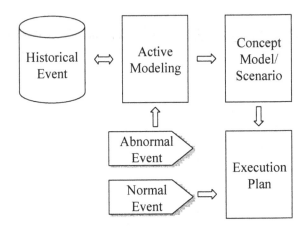

Fig. 1. The basic flow of active modeling

Fig. 1 shows the main idea of active modeling. First, active modeling analyzes historical events into a conceptual model and creates different scenarios for different cases. While a normal event happens, a suitable scenario will be adopted in response. The learning process will be initiated when the current version of conceptual model and scenarios can not properly respond to an abnormal event. After the learning process either incorporates new scenarios or modifies current scenarios, the same (or similar) event would not constitute surprise again in the future.

In order to mitigate damage when a terrorist attack happens, the most important goal of counterterrorism is to detect all of the attacks under plotting before they are launched by terrorists. Typically, a terrorist cell will dynamically change its structure and its members' relationships when someone is arrested, recruit new members, or change attacking targets. Therefore, the continually learning ability of a counterterrorism detection system is very important. Without the learning ability, counterterrorism systems can only detect the threats of the same or similar attacks that have been launched and the attack features that have been identified too. Once terrorist groups plot a new kind of attacks, the system might fail to identify it and the attack incident might become a surprise for all. Even though similar attacks happen again, the system will not be able to raise alarm over the threat. Therefore the continually learning ability is essential to a system that faces a highly dynamic problem domain.

3   Intelligent Terrorism Detection System

The most important issue in counterterrorism-related research is how to detect potential threats in time to prevent terrorist attacks from happening. The Intelligent Terrorism Detection System uses both recognized suspicious activity patterns and abstracted terrorism characters to detect if any potential threat exists in a suspicious terrorist network. Once a threat has been identified, the system will send an alarm to the terrorism analyst to make the final evaluation.

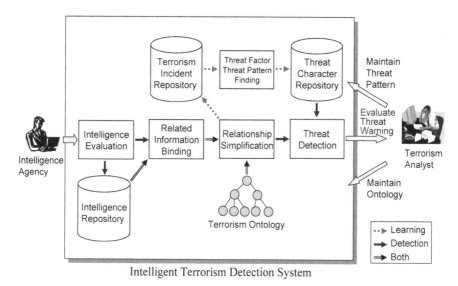

Fig. 2. System architecture of Intelligent Terrorism Detection System

Based on the operation time, the processes of the system can be divided into two phases: the *learning phase* and the *detection phase*, as illustrated in Fig. 2. Firstly, in the learning phase, the system identifies key features from different categories of attacks and identifies activity patterns between terrorists in each terrorist cell from past terrorist incidents. In order to adapt to the dynamic terrorist network and speed up attack threats detection, the activities (for example, repeated activities or fixed-sequence activities) between terrorists will be abstracted to a higher level by the terrorism ontology.

Second, in the detection phase, the system evaluates all intelligence collected by distributed intelligent agencies. If any intelligence is abnormal, the system will retrieve all intelligence related information from the Intelligence Repository and model them into an ARG. Each ARG presents either a normal event or an abnormal event. Then the system will use varied mechanisms to evaluate the ARG. If the event can be recognized, then it is identified as a normal event. When any potential threat has been recognized by Threat Character Library, the system will generate a threat hypothesis to the terrorism analyst.

3.1 Intelligence Evaluation

All intelligence will be evaluated when imported to the system. On the one hand, the system helps the human analyst to filter out large amounts of common intelligence to avoid information overflow. On the other hand, the system can detect threatening intelligence and alert human analysts in time. When new intelligence is fed into the intelligent terrorism detection system, it will be stored in the Intelligence Repository and evaluated by the Intelligence Evaluation process. Once any new intelligence is identified as atypical [7], it would be automatically passed on to the next process, the Related Information Binding process, where related information will be put together for further assessment. For example, if the rule for abnormality is *all fund transfer cannot be larger than $20,000*, the system will send out a warning when a fund transfer exceeds the amount stipulated in the rule.

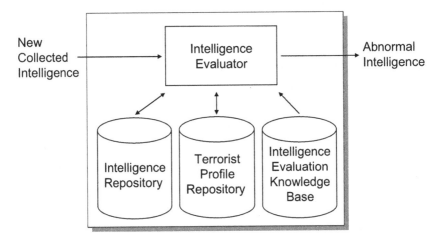

Fig. 3. The operation flow of the Information Evaluation

To speed up the evaluation process, a suspicious group will be checked up only when its new intelligence is abnormal. Fig. 3 shows the detailed operation of the information evaluation process. There are three repositories involved in this operation. The Intelligence Repository stores all imported intelligence. Once the member of a terrorist cell has special skills and material that can be used to launch attacks, how to detect this kind of events is critical for counterterrorism analysis. The Terrorist Profile Repository stores the personal information of all suspicious terrorists. The terrorist profile can help the system keep tracking of each suspicious terrorist. The most important part of this process is the Intelligence Evaluator. It uses both the terrorist profile repository and the Intelligence Evaluation Knowledge Base to assess whether a new intelligence is abnormal or not.

As soon as any new intelligence is detected to be abnormal, the system will retrieve all related intelligence from the intelligence repository to do an advance evaluation. The query language will recursively retrieve all related information from the intelligence repository. Basically, the termination condition of the retrieving operation is based on the small world assumption (the maximum diameter distance between two friends is 6 degrees) or until no more related information can be found from the intelligence repository.

Recognizing abnormal intelligence is a critical step in threat detection. With this operation, the intelligent terrorism detection system can track terrorists' behavior and detect hidden threats from suspicious terrorist networks. In general, the abnormal intelligence is determined by terrorism analysts. The decision has to be made by reference to extensive work experience and domain knowledge. When the Intelligence Evaluation Knowledge Base has been constructed by terrorism knowledge, the Intelligent Evaluator will automatically detect imported abnormal intelligence, as well as prevent information overflow from happening. Both statistics and data mining methods will be used to recognize the features of abnormal intelligence hidden in historical incidents, such as visiting special factories, purchasing explosive material from the black market, or transferring/receiving large amounts of money.

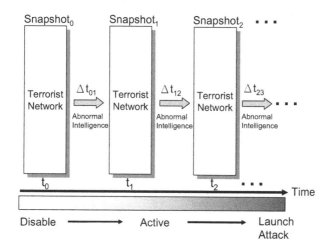

Fig. 4. Based on time sequence, snapshots are used to describe the activity sequence of a terrorist attack

Frequent abnormal intelligence means that the terrorist cell is active and is perhaps preparing to strike anytime. Active modeling can help us illustrate this concept. Following the abnormal intelligence increment, the probability of a terrorist cell to launch attacks is rising too. Initially, the status of a terrorist cell will change from idle to active before launching an attack. When importing collected intelligence into the system, the system will pass normal intelligence and focus on the appearance of abnormal intelligence. In Fig. 4, $Snapshot_i$ represents the status of a terrorist network at time t_i and i will increase along the time axis. The difference between two snapshots is $\Delta_{i,i+1}$, representing that an abnormal intelligence has been detected. Using the sequence of abnormal intelligence can help us track the behavior of terrorists and identify their scheme. The relationship between abnormal intelligence is either dependent or independent. Most of the time, fund transference is the first abnormal intelligence. Money is the most important prerequisite to an attack. Terrorists need considerable funds to purchase weapons and explosives, and to recruit new members.

3.2 Relationship Simplification

Using heuristics can speed up subgraph isomorphism in graph matching, but it is still a time-consuming process at its best. To improve the situation, the intelligence analyst needs to focus only on the essence of the suspicious activity patterns and leave out unnecessary details. Relationship simplification translates a complex graph into a concise one, but keeps all important characters intact.

In this system, a terrorist network is simplified by replacing lower level relations in the terrorism ontology with higher level ones. For example, there are two relations, *Email(A,B)* and *Call(B,A)*, between two suspicious *A* and *B*. These relationships can be abstracted as *Comm(A,B)*, when *Comm* is the superclass of *Email* and *Phone_call* in the terrorism ontology. Basically, the terrorism ontology is a systematic formulation of concepts, definitions, relationships, and rules that captures the semantic content of the terrorist domain.

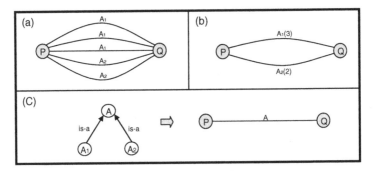

Fig. 5. (a) the original relationship between nodes P and Q, (b) the graph is simplified by gathering the same relations together (c) the graph is simplified by abstracting the same class relations into a higher level

Relationship simplification can be divided into two steps. First, *horizontal relationship simplification* represents same relationships between two nodes by a unique term. For example, in the 911 attack, the primary leaders Mohamed Atta and Marwan Al-Ahehhi had lived together and taken the same airplane flying courses during 1999~2001. We can use *Roommate(times)* and *Classmate(times)* to represent the previous complicated relationships between these two hijackers. The word *"times"* represents the occurrence of a relationship.

Second, *vertical relationship simplification* abstracts different kinds of relationships into a higher semantic level. Even the same relationships can be replaced by one of frequency; the relationship between two entities has many different kinds of relationships between them. In order to reach simplification, we can convert multi-relationships into a higher level term by the domain ontology.

For example, two nodes, *P* and *Q*, have a set of relations between them in Fig. 5(a). The given relation set is $R_1 = \{A_1(P,Q), A_1(P,Q), A_1(P,Q), A_2(P,Q),$

$A_2(P,Q)$}. After the horizontal relationship simplification, the relation set will be changed to $R_2 = \{A_1(P,Q, 3), A_2(P,Q, 2)\}$ as Fig. 5(b). When both relations A_1 and A_2 are subclasses of Class A, the relations between P and Q are abstracted to the higher level relationship A and $R_3 = \{A(P,Q)\}$ as Fig. 5(c).

3.3   Threat Factor and Threat Pattern Finding

In Fig. 6, the potential threat detection will be treated as a feature matching operation. The threat pattern can be identified either by GDM methods or terrorist analysts' definition. Frequent subgraph mining identifies threat patterns from a set of ARGs that represent past terrorist attacks. Term Frequency/Inverse Document Frequency (TFIDF) [10] uses the frequency of terms to form the feature vector of a document (terrorist attack). Based on these vectors, the system clusters them into different groups of terrorist attacks. Then we can recognize the category of a terrorist attack by using the feather of these clusters.

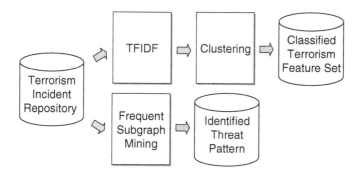

Fig. 6. The operation flow in the learning phase of the system

3.3.1   Threat Factor Finding by TFIDF

TFIDF, an information retrieval approach, weighs terms according to their frequency in each document. Note in this research we replace the documents with terrorist attacks. Each terrorist attack can be represented as a vector,

$$\vec{t} = (t_1, t_2, ..., t_n) \tag{1}$$

$$t_i = TF(tm_i) \cdot IAF(tm_i) \tag{2}$$

$$IAF(tm_i) = \log(\frac{|TA|}{AF(tm_i)}) \tag{3}$$

In Eq(1), each element t_i Eq(2) represents a distinct term tm_i, e.g. a terrorist's skill or a relation between two terrorists, in a terrorist attack. $TF(tm_i)$, term frequency, is the number of times term tm_i occurs in a terrorist attack TA. $AF(tm_i)$, terrorist attack frequency, is the number of terrorist attacks in which term tm_i occurs at least once. $IAF(tm_i)$, inverse terrorist attack frequency Eq(3), can be calculated from the terrorist attack frequency. $|TA|$ is the total number of terrorist attacks.

The tm_i is an important indexing term for a terrorist attack if it occurs frequently in it (the term frequency is high). On the other hand, terms which occur in many terrorist attacks could be rated as less important indexing terms due to their low inverse terrorist attack frequency. The TFIDF can be especially helpful in predicting which category a suspected terrorist group belongs to. Once the category is identified, we can then map the threat patterns of this category onto a suspicious network.

More often than not, the similarity is measured by cosine similarity. The vector angle computes the angle between two vectors to indicate the similarity of these two vectors. The vector angle is defined to be the angle between 0 and 180 degrees that satisfies the relationship

$$\text{Angle}\left(\vec{t_1},\vec{t_2}\right) = \frac{\vec{t_1} \bullet \vec{t_2}}{\left\|\vec{t_1}\right\| \left\|\vec{t_2}\right\|} = \frac{\sum_{i=1}^{n} t_{1i} t_{2i}}{\sqrt{\sum_{i=1}^{n} t_{1i}^{2}} \sqrt{\sum_{i=1}^{n} t_{2i}^{2}}} \tag{4}$$

where vector t_1 and t_2 are the variables containing the n terms of the first and second vectors. If the angle is large, then the similarity is low; if the angle is small, the similarity is high.

3.3.2 Frequent Threat Pattern

Each frequent subgraph found from past incidents is a key factor in a terrorist attack. When frequent subgraphs are identified from the ARGs of terrorist attacks, these subgraphs will be adopted to evaluate whether a suspicious network is planning an attack or not. When each frequent subgraph belongs to a category of terrorist attack, it can help us to uncover the category of any terrorist plot.

In Fig. 7, given an ARG database $GD = \{G_1, G_2, \ldots, G_m\}$, where each graph G_i represents a terrorist network of past attack, when the frequency of G_{si}, a subgraph of some graphs in GD, is higher than a given minimum support, min, G_{si} is a frequent subgraph. After subgraph mining, a set of frequent subgraphs $GD_s = \{G_{s1}, G_{s2}, \ldots, G_{sn}\}$ will be used to detect potential threats in future.

Procedure Frequent_Threat_Pattern_Finding($t_1, t_2, \ldots, t_m$)
begin
   1.   Past attacks $T = \{t_1, t_2, \ldots, t_m\}$
   2.   For $k = 1$ to m
   3. G_k = Relationship_Simplification(t_k)
   4. $GD = GD \cup G_k$
   5. $GD_s = \{G_{s1}, G_{s2}, \ldots, G_{sn}\}$ = Subgraph_Mining(GD)
end

Fig. 7. Algorithm 1 for finding frequent threat patterns

4   Conclusion

Terrorists have continued expressing new techniques to plot new attacks. High technologies as Internet and wireless communication induce them to upgrade their knowledge and to hide their interactions. It is critical and important to predict the behavior and attack plan of terrorist group. The terrorism detection system must have the capability to conquer this challenge effectively and efficiently.

In this paper, we propose an intelligent terrorism detection system which applies active modeling to adapt the conceptual model to solve the highly dynamic problem in counterterrorism by surprise events. It also uses TFIDF to classify the category of potential threats and GDM to recognize the frequent threat patterns from those past incidents. Inexact graph matching is applied to detect the potential threat from suspicious terrorist networks.

Meanwhile, we are evaluating the possibility to modify the terrorism ontology to get a better result in the process of relationship simplification and develop better algorithm to speed up the subgraph isomorphism in the process of subgraph mining. In a foreseeable future, we expect to contribute to the intelligent techniques literatures to prevent further disaster.

Acknowledgments. The research of Yi-Jen Su was partially supported by U.S. National Science Foundation grant: IIS-0326387 and U.S. AFOSR grant: FA9550-05-1-0454.

References

1. Carley, K.: Dynamic Network Analysis. Committee on Human Factors, National Research Council, 133–145 (2003)
2. Krebs, V.: Mapping Networks of Terrorist Cells. Connections 24(3), 43–52 (2002)
3. Chen, P.P., Thalheim, B., Wong, L.Y.: Future Directions of Conceptual Modeling. In: Chen, P.P., Akoka, J., Kangassalu, H., Thalheim, B. (eds.) Conceptual Modeling. LNCS, vol. 1565, pp. 294–308. Springer, Heidelberg (1999)
4. Chen, P.P.: The entity-relationship model: toward a unified view of data. ACM TODS 1(1), 1–36 (1976)
5. Coffman, T., Greenblatt, S., Marcus, S.: Graph-Based Technologies for Intelligence Analysis. CACM 47(3), 45–47 (2004)
6. Cai, D., Shao, Z., He, X., Yan, X., Han, J.: Mining hidden community in heterogeneous social networks. In: Proceeding of Link KDD 2005 (2005)
7. Hollywood, J., Snyder, D., McKay, K., Boon, J.: Out of the Ordinary: Finding Hidden Threats by Analyzing Unusual Behavior. Rand Corporation (2004)
8. Mukherjee, M., Holder, L.B.: Graph-based data mining on social networks. In: Link KDD 2004. Workshop on Link Analysis and Group Detection (2004)
9. Cook, D.J., Holder, L.B.: Graph-based data mining. IEEE Intelligent Systems 15(2), 32–41 (2000)
10. Debole, F., Sebastiani, F.: Supervised term weighting for automated text categorization. In: 18th ACM Symposium on Applied Computing (2003)
11. Foggia, P., Genna, R., nad Vento, M.: Introducing Generalized Attributed Relational Graphs (GARG's) as prototypes of ARG's. In: GbR 1999. 2nd IAPR Workshop on Graph-based Representations (1999)
12. Chen, P.P., Wong, L.Y.: A Proposed Preliminary Framework for Conceptual Modeling of Learning from Surprises. In: Arabnia, H.R., Joshua, R. (eds.) ICAI 2005. Proceedings of the 2005 International Conference on Artificial Intelligence (2005)

To Support Emergency Management by Using Active Modeling: A Case of Hurricane Katrina*

Xin Xu

Computer Science Department
Louisiana State University
Baton Rouge, LA 70803, U.S.A.

Abstract. Reducing the complexity and surprise of emergencies can efficiently alleviate negative impacts on society, either local or global. A good model which can simplify complex events by constructing views to describe changes of entity behaviors and the dynamic and time-varying relationships is the key approach to handle surprise crises. Conventional conceptual modeling has limitations on mapping dynamic real world. Active modeling is suggested as the appropriate way to deal with such surprises. A case study of Hurricane Katrina is used as an example to explain some active modeling concepts including temporal conceptual modeling, multi-perspective modeling, and data, information and knowledge integration.

Keywords: Active Modeling, Emergency, Conceptual Model, Surprise.

1 Introduction

All emergencies including a terrorist attack, natural disaster or other large-scale emergencies are complex and surprise crisis. If the complexity and surprise can be efficiently reduced, an emergency becomes less severe and less disruptive. To better analyze how the different facets of a crisis relate to historical information and current situation, is the first step to reduce surprise and complexity.

However, most of conventional conceptual modeling techniques are only good at describing static side of relationships of entities from a single-perspective way, which are very inadequate to handle such complex and dynamic emergency situations.

Active modeling is a promising approach in analyzing the process of a crisis event, which can describe the important of the real world in a continual way, and help people understand the relationships among entities from different perspectives based on existing knowledge.

The goal of this paper is not on models to handle specific types of crisis. It more focuses on a generic, active-oriented model which applies to all types of emergencies. Hopefully, it may be used in guiding emergencies management, and predicting future.

---

* This research was partially supported by National Science Foundation grant: IIS-0326387 and AFOSR grant: FA9550-05-1-0454.

2 Problems with Current Generic Disaster Process Model

To model an emergency process has been based on analyzing the stages, actions, events, and time frames [1, 2, 3]. A theoretical model is useful in leading to better understanding of the emergency situation, and how a disaster may evolve, which can distinguish between critical elements and noise. The following figure is a generic standard disaster model which is composed of four stages involving *Mitigation, Preparedness, Response, and Recovery*; the underlying principles are *cohesion, communication strategies, and partnerships* [18, 19].

Fig. 1. Circular model of disaster

This linear sequencing model has been subject to criticism. Neal concluded two points [1], firstly, different states occur at the same time, there are no distinguishing divisions; secondly, some surprise events are associated with more than one stage.

This kind of model only helps in understanding the general nature of the disaster process, but provides little insight into real unfolding of an emergency. Rapid change is a main contributing factor in events becoming surprise disasters, and leads to a chaotic response. Chaos is the nonlinear nature of an emergency [5]; this chaos looks like random behavior, however, these unstable behaviors over time stay with clear boundaries. And chaos is also probably a necessary and desirable condition which accommodates an adaptation, cross communications and other such emergent behavior essential to an efficient response [6]. Since conditions are changing constantly, the situations should be assessed and interpreted by available information which matches the context of the situation. Three steps are considered [7]: firstly, getting the right information from overwhelming volumes of data; secondly, determining the course of action, analyzing alternatives, and making decisions; finally, monitoring the execution of instruction and collecting the results as feedbacks for the future decisions.

Neal pointed out that there are many problems in the current uses of disaster periods; the major one is the lack of conceptual clarity for improving scientific and practical use. So he suggested ways the field can recast the use of disaster phases to improve the theoretical and applied dimensions of the field [1].

In order to better dealing with emergency situation, a conceptual model should be capable to analyze surprise scenarios which occur in a crisis, and study abnormal behaviors and changes from different perspectives. This model should also be able to integrate data from historical information (experience, crisis, and surprise), and represent relationship between different degree of importance, and from all aspects of a domain in a dynamic way. The types of model for emergency management will be multi-level and multi-perspective conceptual modeling as well as synchronized, real-time feedback and dynamic constructing.

3 Weakness of Conventional Conceptual Modeling

Most of conventional conceptual modeling techniques primarily emphasize the "static" side of relationships of entities. "Constructs for modeling changes of the entity behaviors and the dynamic and time-varying relationships among them are very inadequate [8]." In the real world, a good conceptual modeling should represent data semantics both related to "what" is important for the application, and related to "when" it is important [14, 15].

The following is an example of a weak conceptual design approach without representing relationship from different perspectives and temporal data. It is well known that there are many resources (police, National Guard, medical team, and fireman) were used during Hurricane Katrina rescue. Based on the normal procedure, local police are responsible for maintaining order of the city during the emergency period. The normal relationship can be simply presented in figure 2.

Fig. 2. Normal relationship between police and citizen

The above single-perspective conceptual modeling may be useful in handling normal disaster situation. However, with unexpected things happened just after storm (dam broken, the city submerged), sixty officers resigned, forty-five were fired, and two committed suicide [13]. The situation became worse and worse, more and more policemen were reluctant to assume their responsibilities. The police force was wracked by desertions and disorganization in Hurricane Katrina's aftermath. The abnormal behaviors lead the above relationship could not keep efficiency during rescue action. We also notice that the change did not occur at a special time point. There was duration, starting from policemen's resignation and ended with insufficient policemen to assume their responsibilities in this disaster.

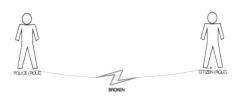

Fig. 3. Broken relationship between police and citizen

Based on the above case, we can see current conventional conceptual modeling techniques are single-perspective, static snapshoot of the real world which are inadequate at continuous information extraction, discovery, and knowledge acquisition. Therefore, a new model should be implemented to support complex applications.

4 Active Modeling

Since there are several limitations of using conventional conceptual modeling techniques to describe the real world, active modeling is a new way to provide a closer conceptualization of reality, which "is a continual process of describing the important and relevant aspects of the real world including the activities and changes under different perspectives based on our knowledge and understanding [8]."

More different characteristics of the world should be integrated in active modeling such as temporal, spatial, psychological, historical, etc. As Peter Chen highlighted, the major need of active modeling is to analyze surprises, crises, and unconventional events such as September-11, Tsunami in Southeast Asia and East Africa, and Hurricane Katrina. Several questions should be answered before using a framework to document and analyze emergency and unexpected crisis [12].

- How do we analyze the surprise/crisis scenarios?
- What information do we need to analyze the surprise/crisis situations?
- What have we learned from the surprises/crises?
- How can we handle surprises/crises in current and future world situations?

Therefore, several specific issues can be regarded as a starting point for active modeling which include capturing human intelligence, visual conceptual modeling, data, information and knowledge integration, self modeling, executable active Entity-Relationship model, uncertainty modeling, temporal conceptual modeling, Multi-perspective modeling, conceptual modeling for multi-agent systems, dynamic reverse modeling, and etc.

The following part will use Hurricane Katrina as a case to explain some proposed concepts of active modeling, and explain how these concepts can help in modeling the continually changing situation.

5 Analyzing Hurricane Katrina by Active Modeling Concepts

5.1 Case Study

Current approaches for predicting and responding to hurricane in emergency management can be described as incorporating weather data into hurricane damage model to determine hurricane intensity, the inner structure of a hurricane, the path of a hurricane [20], and preparing corresponding rescue resources. The main shortages of these methods are focusing on the complexity of hurricane, but overlooking the possibility of potential surprises. The lack of using a multi-perspective and continual process to describe the real world led to the fault of Hurricane Katrina rescue.

Active conceptual model provides a solution to deal with surprises within a complex emergency.

Two issues will be discussed firstly:

- What is surprise and how to deal with surprise?
- How to determine changes which will lead to influence relationships among entities?

For the aspect of surprise, generally, almost every emergency has occurred in the past years. Historical information, experiences have already been stored in various data resources. In normal circumstance, Hurricane Katrina should not be a surprise crisis since people have enough experiences in dealing with such kind of natural disaster. Following standard process, public store food, water, flashlight before Hurricane is coming as what they used to do. So disaster per se is not a surprise either natural or human made. For another example, September-11, from the perspective of public, they though it was a surprise that terrorists hijack planes to attack buildings, but for terrorists, they have spent several years for preparing this attack. Attacking building is one way to make terrorism as well as body bomb and kidnapping. This kind of surprise may be reduced or eliminated before it happened by using some techniques such as data mining, social network analysis. The normal process of "Knowledge Discovery in Databases" which refers to the entire process required extracting knowledge from large-scale databases, and levels of data mining [9], which may be helpful for predicting future event, therefore people will not be surprised about the upcoming emergency.

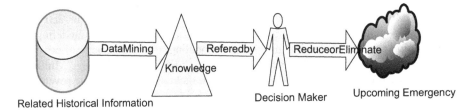

Fig. 4. Normal process to reduce or eliminate surprise before an emergency

The real surprise is the thing that no one has even imaged that it will happen. A more specific explanation of surprise can be defined as *an event that has happened in other circumstances before, but never been considered the possibility of occurring in the upcoming emergency.* For the example of Hurricane Katrina, the dam broken, this surprise resulted in the worst situation. The main reason is people did not take this event into account, so no preparative are arranged even though there are historical experience of dealing with dam broken. This accident induced the New Orleans Police Department (NOPD) spent its first few hours of search and rescue retrieving almost 300 of own officers from rooftops and attics [16], and many officers quit their jobs during the rescue process. The decision maker did not capture the abnormal

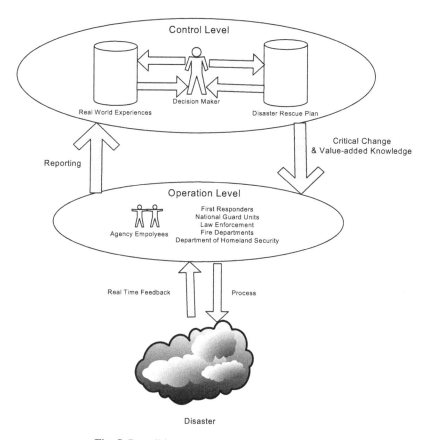

Fig. 5. Describing rescue process by Active model

behavior at the first place, and still use the normal process to handle the operation. The emergency process model is static, could not map the real situation in a dynamic way. For another example, September-11, the surprise is both public and terrorists did not predict that world trade center will collapse. In order to handle the emergency operations dynamically, the following active model should be implemented.

During an emergency environment, event data flow like a steam, entities' behaviors are evaluated at real-time; the relationships can be modified by special triggering action. Current and past experiences are combined together to find a new approach to adapt the new situation.

Determining the changes is the key process to find potential change of relationships among entities, therefore may be useful to prevent future surprise crisis. Multi-perspective and temporal concept can be used to construct a model to describe the changes.

As Bill Watterson says, "I show two versions of reality and each make complete sense to the participant who sees it. I think that's how life works." Multiple Perspectives can be defined as a broad term to encompass multiple and possibly heterogeneous viewpoints, representations and roles, that can be adopted within both a collaborative and non-collaborative context [10, 11].

We can describe a person from a multi-perspective way. People act different roles in the real world. For example, a person works for law enforcement department, his role is a police officer. From another perspective, he has a family; he may have kids, so his second role is father. Generally, the two roles have balanced relationship in a normal situation, and phase is kind of static.

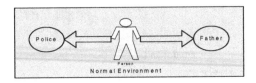

Fig. 6. A person with two roles in normal environment

But when the outside environment changes acutely, the balance may be broken, if decision makers still view the issue from a single-perspective, which will leads to some surprise crisis happens.

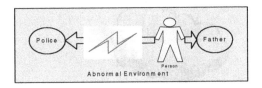

Fig. 7. A person keeps one roles in abnormal environment

In this case, it is obvious that many policemen chose role of father to protect their own home instead to carry out police duty. They resigned job during rescue process which lead the police force wracked, and this result is not considered during a normal disaster management. So it is important to understand the real world in a multi-perspective way.

How to determine changes which will lead to influence relationships among entities is another key issue we should understand clearly. Generally, not all events will cause relationships to be changed. In most cases, things are changing little by little, and only reach a special time point which transnature the relationship.

One approach is to construct a phase plane to represent a relationship among entities. Changes in entities can be plotted over time and used to monitor the status and process. A couple of main attributes of an entity can be set threshold to determine critical changes which will lead to the change of relationship.

For example, we can define the law enforcement as an entity; the number of policemen can be view as a main attribute. The relationship between law enforcement and citizen is maintaining order. We can also set a value of the number of policemen as a threshold to determine whether the law enforcement can work properly.

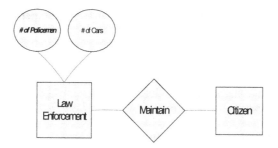

Fig. 8. The relationship between law enforcement and citizen

Actually, a little decrease in the number of policemen is not an abnormal issue, and can not interfere the relationship in figure 8. But if the trend continually occurred, and people do not realize the change, unexpected event will happen.

Monitoring the attribute change and evaluating relationship based time scale is an effective way to describe the real world. An active database is necessary to support the evaluation process by using triggers and event driven. The following figure is using NOPD case to illustrate how to capture and respond to changes. Triggers can be defined by temporal logic [17].

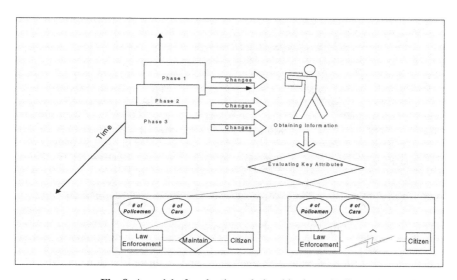

Fig. 9. A model of evaluating relationship dynamically

In this case, an event can be specified as "behaviors of police officers during rescue process including demission, dismissal, disappearance which decreases the total number of policemen." A trigger can be defined as "the number of policemen can work on duty less than 75% percents of the total number of policemen" to inform decision maker that the law enforcement could not work properly as what they are supposed to do. Other triggers can be defined to alert director before the relationship broken between policemen and citizen happens, such as set the percents to 85% in earlier evaluation phase.

5.2 Proposed Approach

Based on the previous case study, a proposed process integrated with supporting technologies is given below which aims to help predict surprises within upcoming natural disasters.

This framework is based on two assumptions:

- All types of natural disasters have already occurred before, and corresponding information, experience, and lessons have been classified and stored.
- People only focus on the major upcoming disaster because of limited resource and the less probability of occurrence of other disasters

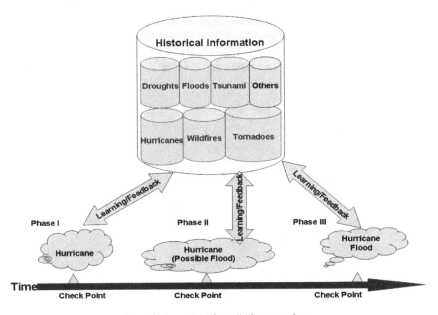

Fig. 10. A model of predicting surprises

In this model, real time information is gathered, and evaluated by domain experts to determine whether previous defined trigger values or events of potential risk are reached. If so, past knowledge will be retrieved from historical information, and combined with current situation to establish new solutions. At the meantime, new learning, lessons are stored for further reference.

An enhanced version of the Entity-Relationship Model is proposed to describe the concept of role from multi-perspective which aims to help determine role changes during emergency management. In this approach, an entity is described by a role attribute, which has multiple values with one weight for each value, and an attribute specifying the time period. The policeman role in Hurricane Katrina can be described as the following figure.

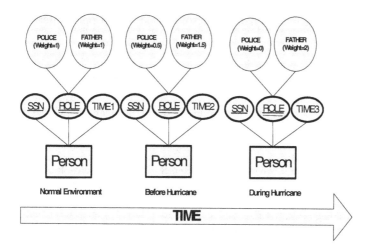

Fig. 11. An example of enhanced ERD

Monitoring the role change of an entity is a good method to predict potential risks which will influence relationships between entities.

To measure the reduction of surprises by using ACM-Learning model, an evaluation model can be used to answer the following questions:

- What potential surprise will happen within current disaster?
- Why the potential surprise will happen?
- When the potential surprise will happen?
- Where the potential surprise will happen?
- How does the potential surprise happen?

The accuracy of describing characteristics of potential surprises which come to occur actually is an approach to evaluate the performance of ACM-Learning.

6 Conclusion

Reducing the complexity and surprise of emergencies is the key to reduce their negative impacts on society. This paper provides an attempt of using active model for handling emergency rescue process. The strength of this model lies in an ability to help decision makers in describing and understanding the relationships among entities and mapping the real world from a multi-perspective way. With this model, efforts at quantification have a base from which to organize the data collected, and also help establish a common base of understanding for all involved.

There are some specific issues need to be future explored:

- How to capture the right information from various changes during an emergency situation
- How to adapt to new situation by learning from past experiences
- How to predict potential changes by integrating current and past knowledge

Active modeling provides a challenging solution to help people better under the complex world.

References

1. Neal, D.N.: Reconsidering the Phases of Disaster. International Journal of Mass Emergencies and Disasters 15(2) (1997)
2. Haas, J.E., Kates, R.W., Bowden, M.J.: Reconstruction Following Disaster. The MIT Press, Cambridge (1977)
3. Frerks, G.E., Kliest, T.J., Kirkby, S.J., Emmel, N.D., O'Keefe, P., Convery, I.: Correspondence. Disasters 19(4) (1995)
4. Kelly, C.: Simplifying disasters: developing a model for complex non-linear events. In: The Disaster Management: Crisis and opportunity: Hazard Management and Disaster Preparedness in Australasia and the Pacific Region Conference, Cairns, Queensland (1998)
5. Keil, L.D.: Chaos Theory and Disaster Response Management: Lessons for Managing Periods of Extreme Instability. In: Koehler, G.A. (ed.) What Disaster Response Management Can Learn from Chaos Theory, California Research Bureau, California State Library, Sacramento (1996)
6. Priesmeyer, H.R., Cole, E.: Nonlinear Analysis of Disaster Response Data. In: Koehler, G.A. (ed.) What Disaster Response Management Can Learn from Chaos Theory, California Research Bureau, California State Library, Sacramento (1996)
7. Chen, P.P., Wong, L.Y.: A Proposed Preliminary Framework for Conceptual Modeling of Learning from Surprises. In: Arabnia, H.R., Joshua, R. (eds.) ICAI 2005. Proceedings of the 2005 International Conference on Artificial Intelligence, Las Vegas, Nevada, USA, June 27-30, 2005, vol. 2, pp. 905–910. CSREA Press (2005)
8. Chen, P.P., Thalheim, B., Wong, L.Y.: Future Directions of Conceptual Modeling. In: Chen, P.P., Akoka, J., Kangassalu, H., Thalheim, B. (eds.) Conceptual Modeling. LNCS, vol. 1565, pp. 294–308. Springer, Heidelberg (1999)
9. Fayyad, U., Piatetsky, G., Smyth, P.: From Data Mining to Knowledge Discovery in Databases. AAAI, Providence, Rhode Island (1997)
10. Park, K., Kapoor, A., Scharver, C., Leigh, J.: Exploiting Multiple Perspectives in Tele-Immersion. In: The Proceedings of IPT 2000, June 19-20, 2000, Ames, Iowa, CDROM (2000)
11. Park, K., Kapoor, A., Leigh, J.: Lessons Learned from Employing Multiple Perspectives In a Collaborative Virtual Environment for Visualizing Scientific Data. In: The Proceedings of ACM CVE 2000, September 10-12, 2000, San Francisco, CA, pp. 73–82 (2000)
12. Chen, P.P.: Active Conceptual Modeling: The New Frontier in Research and Development. In: Active Conceptual Modeling of Learning Workshop, San Diego, California (2006)
13. Anderson, W.: This Isn't Representative of Our Department: Lessons from Hurricane Katrina for Police Disaster Response Planning, Disasters & the Law, Professor Daniel Farber (2006)
14. Gregersen, H., Jensen, C.S.: Temporal Entity-Relationship Models-A Survey. IEEE Transactions on Knowledge and Data Engineering 11(3), 464–497 (1999)
15. Khatri, V., Ram, S., Snodgrass, R.T., Vessey, I.: Strong vs. Weak Approaches to Conceptual Design: The Case of, Temporal Data Semantics. In: Working Paper Series

16. Katrina, H.: The Role of the Governors in Managing a Catastrophe: Hearing Before the S. Comm. On Homeland Security and Governmental Affairs, 109th Cong. 3 (testimony of Warren J. Riley, Superintendent of the New Orleans Police Department) (2006)
17. Sistla, P., Wolfson, O.: Temporal Triggers in Active Databases. IEEE Transactions on Knowledge and Data Engineering (TKDE), 471–486 (1995)
18. http://en.wikipedia.org/wiki/Emergency_management (2006/12/08)
19. Doucet, M.E.: Overview of Canada's Emergency Management Framework, Office of Emergency Management. In: CFIA (2004)
20. http://svs.gsfc.nasa.gov/stories/hurricanes/index.html (2006/12/08)

Using Ontological Modeling in a Context-Aware Summarization System to Adapt Text for Mobile Devices

Luís Fernando Fortes Garcia[1,3], José Valdeni de Lima[2], Stanley Loh[1,4], and José Palazzo Moreira de Oliveira[2]

[1] Lutheran University of Brazil, Department of Computer Science, Av. Farroupilha, 8001, Canoas - RS, Brazil - 92425-900
[2] Federal University of Rio Grande do Sul, Institute of Informatics, Av. Bento Gonçalves, 9500 Porto Alegre – RS, Brazil, 91501-970
[3] Faculdade Dom Bosco de Porto Alegre, Department of Computer Science, Rua Marechal José Inácio da Silva, 355, Porto Alegre – RS, Brazil – 90520-280
[4] Catholic University of Pelotas, PPGINF, R. Félix da Cunha, 412 Pelotas - RS, Brazil - 96010-000
luis@garcia.pro.br, {valdeni,palazzo}@inf.ufrgs.br, sloh@terra.com.br

Abstract. This paper presents a context-aware text summarizer based on ontologies intended to be used for adapting information to mobile devices. The system generates summaries from texts according to the profile of the user and the context where he/she is at the moment. Context is determined by spatial and temporal localization. Ontologies are used to allow identifying which parts of the texts are related to the user's profile and to the context. Ontologies are structured as hierarchies of concepts and concepts are represented by keywords with a weight associated.

Keywords: context-aware summarization, ontologies, adaptation, mobile computing.

1 Introduction

Nowadays, people are seeking for casual learning in academic and business organizations. Casual learning differs from formal courses because apprentices look themselves for content and teaching material without having to register or even pay for that. With the growing use of electronic media, apprentices can obtain qualified content in websites, digital libraries and even in CD or DVD materials. However the huge amount of information available electronically all over the world and in different media causes the information overload and thus makes difficult the learning process.

One alternative to minimize the overload is the use of automatic text summaries generated from original documents. This contributes to the reduction of information overload, since only a selection of the most relevant content is exhibited and people can better select information sources where to deepen.

A special case of text summarization is due to mobile computing. Mobile devices such as cellular phones, handhelds, smart phones, pocketPCs, palmtops and some

P.P. Chen and L.Y. Wong (Eds.): ACM-L 2006, LNCS 4512, pp. 144–154, 2007.
© Springer-Verlag Berlin Heidelberg 2007

notebooks have limitations in display size, storage and processing. When people use these devices, they need summarized information that maintains the relevancy and quality of the information as in its original sources but that regards the constraints of the mobile device.

Other problem with traditional text summarization is that results are produced independent of the context of the user. In the majority of the summarization works, the same summary is generated for different people. Summarizers do not consider the different interests that people have. Furthermore, traditional summarizers do not hold the possibility of people using devices in different situations. The same person may play different roles and then he/she may have different interests depending on the moment. The combination of a person and a situation defines a specific scenario.

For these reasons, text summarization should concern questions related to where the user is and what is he/she interested on in order to appropriately generate the summary. This situation is known as "context-aware computing". Schilit & Theimer (1994) use the term to represent an emerging area to investigate applications that adapt themselves according to the user's location, to a group of people, to object near the user or to changes occurred with the objects along the time. These conditions compose the context of the application or of the user. According to Chen (2005), "Context is a description of the situation and the environment a device or a user is in".

Schmidt et al. (1998) describe 6 factors that influence the context: the information about the user, the social environment, the user's tasks, the location, the infrastructure and the physical conditions. According to Dey (2001), there are 4 kinds of context: the computational context, the user context (related to the social situation of the user), the physical context (related to the environment conditions) and the temporal context (date, time and seasons).

Systems and devices have to adapt to the context, that is, they must alter their behavior (internal tasks and communication interfaces) according to changes occurred in the context (Henricksen et al., 2002). In this sense, summarizers must consider the context to elaborate an extract from a text, generating a different summary for each user and context, according to the target user's interest and the conditions where he/she is at the moment.

This paper presents a system that generates context-aware text summaries for mobile devices users, modeling the context of the user (interest and situation) with the help of domain ontologies. In this paper, context-aware summarization is defined as the use of additional information, associated with spatial localization and user's interest, in the method for determining the relevance of extracts to be used for generating a text summary. Domain ontologies are employed to describe and represent the user's interest and the user's situation (localization or moment), so that the summarizer is able to generate a different summary according to the combination user + situation. This combination (a user plus a situation) is considered a unique context.

The paper is structured as follows. Section 2 presents related works. Section 3 discusses the theoretical aspects of the modeling and the structure of the domain ontology utilized in the context-aware summarization system. Subsections 3.1 and 3.2 present details of the system (architecture and the summarization process). Section 4 presents concluding remarks.

2 Related Work

The adaptation of texts for mobile devices was discussed in many works. Gomes (2001) proposes heuristics that make possible to display long documents in size-limited devices, with no damage to content understanding. Corston-Oliver (2001) presents an approach to the adaptation of texts that must be displayed in small devices, based on text compaction, generating a telegraphic representation of each sentence by excluding some elements. Buykkokten's proposal (2001 and 2002) for text summarization is implemented in five methods, in which each web page is split in semantic textual units that can be partially displayed. The user can explore successive portions of text in different levels, according to his/her particular needs. The adaptation developed by McKeown (2001), McKeown et al (2003) and Muñoz (2003) takes into account the user's profile, when adapting contents to mobile devices. However, summarization does not consider contextual information, since it is limited to only consider the user's profile (patient or physician) to generate different summaries.

Recently ontologies are being used to structure information. An ontology is a kind of knowledge about the domain and can help in representing the user's profile. For example, Middleton et al. (2003) use a topic ontology to represent interests of users. Topics are extracted from CORA, a domain ontology about Computer Science. Documents are classified in topics by using the k-NN method; the topics associated to the user are those related to documents browsed by users. The work also uses the hierarchy of topics to infer new topics of interest.

Gauch et al. (2003) also use concepts from an ontology to represent profiles. The web pages visited by the user are automatically classified into concepts of an ontology. Profiles are generated from the analysis of the behavior of the user, specifically the content, length and time spent on each visited page. The ontology is based on subject hierarchies from Yahoo, Magellan, Lycos and the Open Directory Project. Only top concepts in the hierarchy were used, leading to more general areas of a user's interest.

The missing point in the cited works is that none of them combines text summarization, profiles, ontologies and context-awareness. The goal of the present paper is to discuss the use of ontologies for representing context (user's profile and situations) in a context-aware text summarizer.

3 Ontologies for Context-Aware Summarization

Users may have different interests or preferences according to conditions where they are. For example, when someone is physically present at the stock exchange, we could consider natural that he/she was provided with summaries about the financial market. However, when he/she goes to a shopping mall, the new context changes the summarization focus to other issues concerning products, sales and movies.

Contextual information is useful to indicate "what" should be delivered to the user, "where" and "when". The user's profile determines which information is interesting for the user. However, depending on where the user is and when (what moment or time), the interest may be biased to a different subset. Thus, information delivered to user depends on 2 sets of features: information about the user's profile and information about the user's condition at the moment (local and time).

In this work, profiles are persistent interests. We believe that interests represented in the profile remains stable for a short time. Contexts are related to physical or geographical places (spatial localization) and to special conditions of the user at a moment (temporal localization). It is interesting to note that the same context may occur in different places; for example, a context defined as "vacations" may occur in a beach house or in a mountain hotel. Contexts are also determined by temporal information. For example, "beginning of the day" may be a context that may occur at a specified time during the week (6:00 pm) or in an interval in the weekend (between 9 and 10 pm). This context may used to deliver news for the user. Note that this context may be independent of the physical place ("deliver news to user in that time wherever he/she is") or may be dependent ("only deliver news in that time if user is in the kitchen").

As context changes, the same user may receive different information according to different situations. For example, in an "office" context, user should receive financial news about the market and, while in "vacations", he/she will receive general news.

The system presented in this paper employs information about the user's profile and information about the current context of the user (temporal and/or spatial localization) to determine the kind of text summary that will be generated to him/her. To represent the different kinds of profiles and contexts, this work uses domain ontologies. A domain ontology is a description of "things" that exist or can exist in a domain, Sowa (2000) and contains the vocabulary related to the domain, Guarino (1998). The domain ontology determines which concepts of the real world are being considered and describes how they can be detected. In this paper, domain ontologies are structured as concept hierarchies with each concept being described by keywords. Associated to each word, there is a weight that determines how much the word identifies the concept (profile or context). These keywords will be used to determine which parts of the text will be selected for composing the summary to be delivered to the user. Keywords work like text filters; only phrases that contain one of the keywords are considered relevant to the user (this is an adaptation of the "cue-phrases" concept by Paice (1981) As we explain in the next section, the summarization technique evaluates the relevance of each phrase according to the presence of keywords (profile words and contextual words).

The ontology used for profiles is based on categories about general knowledge, following suggestion of Gauch et al. (2003) and Labrou and Finin (1999) that use categories from *Yahoo!* as an ontology to describe content and features of items. Only high level areas (top topics) are being considered at the moment. Keywords associated to each concept in the ontology and their weights were defined by experts in each area a. Table 1 presents part of the ontology employed in the current system for describing profiles, including some terms that identify each profile. The ontology for describing contexts was created manually by the authors. The definition of terms and weights was made by the authors. Table 2 presents part of the ontology employed in the current system for describing contexts (temporal or/and spatial localization). In table 2, some terms employed to identify each situation are shown. These examples do not show the complete ontologies; only part of the concepts is showed and only a few terms were listed for example purpose.

Table 1. Part of the profile ontology

Profile	Keywords
Architecture	house, flat, furniture, decoration, renovate, architecture, curtain
Education	Student, teacher, professor, learning, teaching, school, university
Computers	Computer, software, hardware, cpu, internet, modem, web, ADSL, virus
Fashion	Boot, t-shirt, pants, shoe, coat, tie, fashion, model, clothes, wear
Health	Physician, patient, disease, diagnosis, ambulance, nurse, hospital, clinic, medication, surgery, diet

Table 2. Part of the context ontology

Situation	Keywords
Office, business	deadline, delay, schedule, chief, officer, office, business, technology, dolar, stock, wall street, task
Shopping	buy, shop, delivery, liquidation, discount, sell, sold out, credit card
Vacations, Day off	Vacations, rest, summer, beach, day off, holyday, weekend, entertainment
Home	Television, TV, rest, home, dinner, kitchen, cuisine, talk show

At the moment, the ontologies embedded in the summarizer do not include inference rules. However, as the goal of this paper is also to propose an ontological modeling of context, we will discuss this issue now. One way to extend the ontology is incorporating inference rules over the structure of concepts. The profile ontology may accommodate rules about cross interest. For example, if a user likes "text mining" topic, it is likely that he/she will be interested in issues related to "data mining".

In the context ontology, we can have rules about the combination of situations. For example, we can define that "summer vacations" is an aggregation of the context "summer" with the context "vacations". In this case, the aggregated concept (the new context) inherits the terms from both child concepts. The process is similar to the behavior of object-oriented models. In another case, someone can define that context X is incompatible with context Y (for example, "work" and "leisure"). In this case, information about one context may restrict the information about the other context and intersections must be disregarded. In the given example, if the user is in a "work" context, the summarizer must eliminate whatever information about "leisure".

A special case of inference rule is the combination of the two ontologies (profiles and contexts). Information about one concept in one ontology may be used to infer concepts in the other ontology. For example, if the user has in his/her profile that "he/she is only interested in books", we can infer that he/she will never be in the "cinema" context. Similarly, we can define a rule such that if the user is habitually in the "cinema" context, then he/she should have "movies" associated to his/her profile.

3.1 Architecture of the System

Basically, the system receives information from a mobile device, including the user identification and the context where the user is at the moment. Then the system retrieves contextual words according to the user's profile (the user's profile is stored in a database of profiles) and words associated to the identified context (according to the

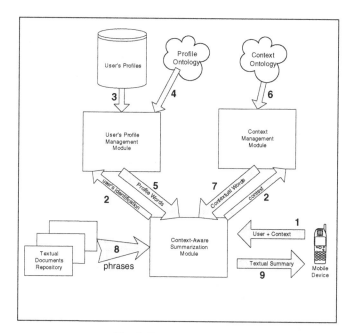

Fig. 1. System architecture

context profile). The final task is to select phrases from textual documents stored in the repository. Each document is processed at one time. The process of filtering phrases is discussed in the next section. For each document, one summary is elaborated and delivered to the user's mobile device. We assume that there is a software module in the mobile device that is responsible for showing the textual summary at the device display.

The system consists of three main modules: (i) User's Profile Management Module; (ii) Context Management Module and (iii) Context-Aware Summarization Module (fig.1). The information flow identified in the figure is described below:

1. Receives information about the user and his/her context
2. Informs User's Profile Management Module and Context Management Module
3. User's Profile Management Module identifies the user's interest in his/her profile
4. Profile Management Module retrieves keywords associated to the corresponding profiles
5. Profile Management Module returns keywords and weights
6. Context Management Module retrieves keywords associated to the corresponding context
7. Context Management Module returns keywords and weights
8. Summarization module separates phrases from the text and ranks the phrases according to the relevance formulae
9. Summarization module sends a summary to the user

The mobile device, where the summarized information will be displayed, is responsible for obtaining information related to spatial localization, temporality and user's identification. For space limit reasons, this paper will not discuss how this information is identified by the mobile device or with the help of other device. We assume that there is a device (hardware or software) that identifies the context (local and time) where the user is in. We can assume that some kind of hardware or software is able to recognize the user by its device and, using a geographical information system, it can determine the place where the user is. Local time may also be informed as contextual information.

The document repository contains the textual documents that will be summarized with the context-aware approach. For each document in the repository, a summary is constructed. Documents could be previously selected and only some of them could be analyzed, but we do not discuss this in this paper.

The database with users' profiles stores information about users and the profiles associated to them (one or more). A profile represents the user's interest in terms of subjects or themes. The user's profile is dynamic and is constantly updated from information about the user. This paper will not discuss how the profiles are created and maintained, but reader can imagine a module that collects explicitly information from the user through questionnaires, forms or interviews or that observes the user's actions in other systems and defines the profile that best match the user's behavior. An extension of this model could include other information as mood, interface preferences and cognitive data.

The profile ontology helps in identifying words related to each profile. These words will be used by the summarization module jointly with context words. The context ontology has definitions about which contexts exist and also the terms related to each context. These context words are used in the summarization process. The user's profile management module receives the user's identification and searches in the profile database for profiles associated to the user. Then, it retrieves from the profile ontology the words associated with the respective weights.

The context manager module is responsible for retrieving contextual words from the context ontology, based on the context identified by the mobile device. The words and their respective weights are then passed to the summarization module.

The context-aware summarization module implements an automatic text summarization technique improved with the inclusion of contextual words supplied by the user's profile management module and by the context management module. The summarization process is detailed in the next section.

The architecture is open, so the summarization module can be implemented and/or expanded with different summarization techniques and with the addition of new kinds of contextual information.

3.2 The Summarization Process

The summarization process starts by receiving contextual words (from the Context Management Module) and profile words (from the User's Profile Management Module) and then picks each document from the Document Repository to generate a summary for each document.

The context-aware summarization approach is based on the idea that the presence of some words in a text demonstrates or are clues that this text is about certain theme or subject. This is the fundament of extractive summarization techniques that select sentences from the original text to compose the summary. The process of selecting sentences evaluates the relevance of each phrase by the relevance of the words that compose it. The weight of each word may be determined by different methods as for example the frequency of the word in the text, the presence in titles and subtitles, etc.

The process proposed in this paper increments the value of sentences extracted from traditional methods of summarization by adding relevance values related to the user's profile and his/her context. As a given word tends to be more significant or frequent in a certain context than in others, its presence in a sentence may increase the relevance of this sentence for users interested in that context. This makes some parts of the text more relevant than others.

It is important to remember that the selection of contextual words and profile words are made previously. Profile words are those related to the profile associated to the user in question, according to the database of profiles and to the profile ontology. Contextual words are those associated to the identified context, according to the context ontology.

The context-aware summarization technique, that calculates the relevance of a sentence, was created based on the TFISF algorithm Larocca Neto (2000). This algorithm was adapted for the present study to incorporate contextual information. The original TFISF algorithm calculates the importance of a word w in a sentence s by the following formulae (1):

$$TFISF(w,s) = TF(w,s) \times ISF(w) \tag{1}$$

where TF(w,s) is the number of times the word w occurs in the sentence s, and ISF(w) is the inverse frequency, calculated with the following formulae (2):

$$ISF(w) = \log(\frac{tam(s)}{SF(w)}) \tag{2}$$

where SF(w) is the number of sentences in which the word w occurs, and tam(s) is the total number of words that compose the sentence s.

The final calculus of the relevance value of the sentence s with the addition of context-awareness is shown in the following formulae (3):

$$TFISFca(s) = \sum_{j=1}^{n} \left[TFISF(w_j, s) \times IP(w_j) \times IC(w_j) \right] \tag{3}$$

where TFISFca(s) is the relevance value for the sentence s according to the context-aware technique presented in this paper; and TFISF(w_j,s) is the relevance of the word j in the sentence s according to the original TFISF algorithm; and IP(w_j) is the weight of the word j given by the profile ontology; and IC(w_j) is the weight of the word j given by the context ontology; being n the number of significant words in the sentence s.

If the word j is not present in the profile of the user, the value for IP(w_j) is 1. If the word j is not present in the current context, the value for IC(w_j) is 1. The weights of the words in the ontologies range from 2 to 10. For the moment, we are assuming one unique profile as valid at a moment.

In this technique, besides the traditional TFISF calculus for relevance of sentences, a sentence that contains profile or contextual words have its relevance value multiplied by the weights associated to these words in the ontologies, increasing the chances of that sentence to participate in the final summary.

If one word in a sentence is not in the user's profile or in the current context, its original relevance value is maintained and its original contribution for the sentence selection is preserved.

The process excludes words known as stopwords, such as prepositions and articles. Thus, only significant words are considered in the relevance calculus.

The result of the process (the textual summary) is composed by the most relevant sentences, those with the highest relevance values, after the calculus is made for all sentences in the textual document.

4   Conclusions

Experiments carried out with the system in a real situation. Texts were gathered from a Brazilian periodic (Veja). Ontologies were manually elaborated. The profile ontology has 10 profiles (economy, fashion, sports, politics, medicine, religion, turism, police, arts, education) while the context ontology was created with 5 contexts (home, shopping, vacations, work, supermarket). Approximately 30 words were associated to each concept in the ontologies. Summaries were generated by the proposed method for each combination of one profile and one context. Human judges rated each summary according to the combination, analyzing relevance and coherence. The results indicate that 67% of the summaries were rated as relevant to the combination of profile+context and 80% of summaries were judged as coherent to the original text by maintaining information coherent.

Although the promising results, we must carry out another experiment with an objective evaluation. The subjective evaluation is important but suitable to mistakes. The problem is to find a measure appropriated for the purpose (personalized and context-aware summarization of texts).

During the work, we found some obstacles. The elaboration of the ontologies by hand is time-consuming and error-prone. The definition of concepts for each ontology is theme of an other paper. We are investigating the use of machine learning methods to semi-automatically identify concepts determine. Another problem is the definition of words and their weights for each concept. We intend to use text mining methods such as TFISF, Latent Semantic Indexing and Bayesian networks to automatically find words and weights.

Other difficulties related to the ontologies include the ambiguity problem (when a keyword appears in more than one concept, generating summaries out of the context) and the lack of a stemming method for preprocessing of the words.

A special case for future works is the study of the coherence among the phrases that compose a summary. In the current stage, summaries are generated by the concatenation of the selected phrases. This can generate coherence problems when passing from one phrase to the next.

In future works, context must be determined by automatic mechanisms and profiles will be learned by intelligent software tools.

The main contributions of this work are the development of an open architecture for context-aware systems and the automatic text summarization process based on context and personalized information. Results from the experiment lead us to the conclusion that the proposed method is promising and can provide more focused information.

The application of the proposed method in learning environments and situations is well appropriated since apprentices can receive personalized and summarized information wherever they are. This can reduce the time for knowledge acquisition and minimize the overload problem. Summaries are oriented to interesting themes and momentary situations, freeing humans from the need for selecting text parts and from giving information about their interests.

References

1. Anderson, C., Domingos, P., Weld, D.: Web Site Personalizers for Mobile Devices. In: International Joint Conference on Artificial Intelligence, Proceedings... [S.l.:s.n.], vol. 17 (2001)
2. Buykkokten, O., et al.: Efficient Web Browsing on Handheld Devices Using Page and Form Summarization. ACM Transactions on Information Systems, New York 20(1) (January 2002)
3. Chen, A.: Context-Aware Collaborative Filtering Systems: Predicting the User's Preferences in Ubiquitous Computing. In: CHI 2005. Proceedings Conference on Human Factors in Computing Systems, pp. 1110–1111 (2005)
4. Corston-Oliver, S.: Text Compaction for Display on Very Small Screens. In: Meeting of the North American Chapter of the Association for Computational Linguistics, Proceedings... [S.l.:s.n.], vol. 2 (2001)
5. Dey, A.: Understanding and Using Context. Personal and Ubiquitous Computing, London 5(1), 4–7 (2001)
6. Gauch, S., Chaffee, J., Pretschner, A.: Ontology-based personalized search and browsing. Web Intelligence and Agent System 1(3-4), 219–234 (2003)
7. Gomes, P., et al.: Web Clipping: Compression Heuristics for Displaying Text on a PDA. In: MOBILE HCI, 2001. Proceedings.. [S.l.:s.n.] (2001)
8. Guarino, N.: Formal Ontology and Information Systems. In: FOIS 1998. International Conference on Formal Ontologies in Information Systems, Trento, Italy, pp. 3–15 (1998)
9. Henricksen, K., Indulska, J., Rakotonorainy, A.: Modelling Context Information in Pervasive Computing Systems. In: Mattern, F., Naghshineh, M. (eds.) Pervasive Computing. LNCS, vol. 2414, Springer, Heidelberg (2002)
10. Labrou, Y., Finin, T.: Yahoo! as an Ontology - using Yahoo! categories to describe documents. In: CIKM 1999. 8th International Conference on Knowledge and Information Management, Kansas City, MO, pp. 180–187 (October 1999)
11. Larocca Neto, J., Santos, A.D., Kaestner, C.A.A., Freitas, A.A.: Document clustering and text summarization. In: Proc. 4th Int. Conf. Practical Applications of Knowledge Discovery and Data Mining, pp. 41–55 (2000)
12. McKeown, K.: PERSIVAL, a System for Personalized Search and Summarization over Multimedia Healthcare Information. In: JCDL. Proceedings.. [S.l.:s.n.] (2001)

13. McKeown, K., Elhadad, N., Hatzivassiloglou, V.: Leveraging a Common Representation for Personalized Search and Summarization in a Medical Digital Library. In: Joint Conference on Digital Libraries, 2003. Proceedings.. [S.l.:s.n.] (2003)
14. Middleton, S.E., Shadbolt, N.R., Roure, D.C.D.: Capturing interest through inference and visualization: ontological user profiling in recommender systems. In: KCAP 2003. International Conference on Knowledge Capture, pp. 62–69. ACM Press, New York (2003)
15. Muñoz, M., et al.: Context-Aware Mobile Communication in Hospitals. IEEE Computer, [S.l.] 36(9) (September 2003)
16. Paice, C.: The automatic generation of literature abstracts: an approach based on the identification of self-indicating phrases. In: Oddy, R., et al. (eds.) Information Retrieval Research, London: [s.n.] (1981)
17. Pardo, T., Rino, L.: TeMário: Um Corpus para Sumarização Automática de Textos. São Carlos: Universidade de São Carlos, Relatório Técnico (2003)
18. Schilit, B., Theimer, M.: Disseminating Active Map Information to Mobile Hosts. IEEE Network, [S.l.] 8(5), 22–32 (1994)
19. Schmidt, A., Beigl, M., Gellersen, H.: There is more to Context than Location. In: International Workshop on Interactive Applications of Mobile Computing, 1998. Proceedings.. [S.l.:s.n.] (1998)
20. Sowa, J.F.: Knowledge representation: logical, philosophical, and computational foundations. Brooks/Cole Publishing Co., Pacific Grove, CA (2000)

Accommodating Streams to Support Active Conceptual Modeling of Learning from Surprises

Subhasish Mazumdar

Department of Computer Science
New Mexico Institute of Mining and Technology
Socorro, NM 87801, USA
`mazumdar@nmt.edu`

Abstract. We argue that a key requirement on an information system that can implement an active conceptual model of learning from surprises is the ability to query data that is not query-able by content, especially data streams;we suggest that such data be queried by context. We propose an enhancement of entity-relationship modeling with *active* constructs in order to permit such streams to have context-based relationships with standard data. We propose a framework wherein the analysis of surprises and the subsequent monitoring of states that are ripe for such events are possible by the use of such contexts.

1 Introduction

The aim of active conceptual modeling is to provide "a closer conceptualization of reality, ... a multilevel and multi-perspective abstraction" [1]. One benefit is that it would enable information technology to act in synergy with the human process of learning from experience. In particular, sudden deviations from the normal course of events, or *surprises* [2], are crucial in such learning because they indicate that the predictive model that we use routinely, to provide us with normalcy and stability in coping with a dynamic complex world, needs modification; analysis of the causes of the surprise can lead to an improved model.

Owing to a closer capture of reality, implementations of information systems based on active conceptual modeling of learning will allow us to revisit, identify, and analyze past surprises, validating or uncovering cause-effect relationships among events; these could be used to infer at some time in the future that the conditions then are ripe for an imminent surprise and thus help avoid or mitigate (or exploit) its disruptive consequences[1]. Laying out the causal sequence of events and their ramifications, we can look for patterns and generalize from them, training either automated pattern analyzers, or human experts, or both.

Conventional static conceptual modeling is typically used to isolate an appropriate fragment of the universe of discourse for which discrete snapshots

---

[1] Until our prediction model is perfect, we can never predict all future surprises; however, even *some* successful predictions can be beneficial.

P.P. Chen and L.Y. Wong (Eds.): ACM-L 2006, LNCS 4512, pp. 155–167, 2007.
© Springer-Verlag Berlin Heidelberg 2007

are stored in a database that is kept as current as possible through frequent updates under the control of a database management system which maintains consistency and facilitates querying. To extend this approach and accommodate active conceptual modeling, one may conclude that all we need to do is store every database snapshot explicitly or implicitly (store enough to be able to create them dynamically), capture the changes themselves plus the relationships between or among those changes. But such an extension is not straightforward for a number of reasons.

First, database updates are typically not done in real-time. Second, database management systems rearrange update transactions in an unpredictable order conforming to serializability, not necessarily the serial order of submissions; further, transactions can be aborted owing to unpredictable system conditions, such as deadlock, and be re-executed later. For these two reasons, the order of database snapshots do not necessarily reflect the order of events and system-generated timestamps on data do not reflect the occurrence time of events. Explicit storage of event time is not a simple solution either as synchronization between distinct autonomous databases is problematic. All this exacerbates the cause-effect inference, especially when different databases are linked. Third, unless events underlying the transactions are themselves stored, they may be impossible to infer from a sequence of snapshots owing to the many-to-one map from events to states.

Fourth, and most important here, update transactions are discrete events chosen by a designer with a certain model of the miniworld. An analysis of a surprise may require a snapshot involving *unforeseen* entities and/or a state of the world between updates. For example, when someone assaults and steals from a customer withdrawing money from a bank's automatic teller machine, the physical appearance of the criminal and his vehicle may turn out to be much more important than the details of the banking transaction of the customer. Fifth, of all recordable data in the planet, the data that is query-able by content (QBC) (representable in ASCII) is much less than — by several orders of magnitude — the opposite kind: image, video, and sound: music, and telephony (which constitutes the bulk) [3]. The latter kind of data mainly takes the form of streams. Stream data can be both non-QBC (as above) or QBC (e.g., sensor data). Many data streams cannot be stored as discrete snapshots in a database as they are too voluminous and do not correspond to events or transactions. Current technological developments indicate that such data will grow; for example, a person will be able to wear tiny digital devices such as microphones and cameras and essentially record his/her life, the motive being the recall of past events, both pre-planned and unexpected. The proliferation of video cameras in cellphones, at traffic intersections, in cars, and at public places will create massive distributed data banks of such stream data that can potentially be linked to trace a large number of events.

From the fourth and fifth point above, we conclude that when we will need to query unforeseen attributes and entities and/or states between database updates, it is quite likely that we will have to make use of non-QBC data streams. We will have to access such stored data, play it back for the duration of a surprise event, perhaps plus preamble and sequel; and for causal analysis, access other related data in database snapshots and through them, possibly other non-QBC data. Our approach to data that is *not* query-able *by content* is to query it *by context* which, in this case, means associations with snapshots in databases.

The primary observation behind our approach is that a surprise, if it is of any importance, must necessarily encompass data that we have *not* hitherto considered to be important enough to include in our model and cover under the gamut of a database kept updated and consistent. To analyze the surprises, therefore, we have to go back and *piece together all available data that may shed light* on the actions that led to the surprise. It is inevitable that a large amount of such data will be non-QBC data. Hence our attempt aims at facilitating this post-facto analysis.

In general, these other non-QBC data may come from outside the enterprise. Hence, our framework provides for them to be linked using contextual clues. Modeling such sources as active entities allow a (perhaps semi-automated) process of linking various sources.

This paper is structured as follows. First, we introduce the proposed extensions to the ER model and outline our framework for learning from surprises. Next, we take a closer look at an application involving a classroom. Subsequently, we review related work and conclude with remarks about our overall approach.

2   Our Approach

In this section, we will introduce active ER modeling constructs, and use them to outline our approach for using contextual clues in linking with streams.

2.1   Active ER Model

We first propose an extension of the ER model with two constructs: active attributes and active entities. Intuitively, an active attribute is a generalization of the notion of the derived attribute. Recall that a derived attribute of an entity is one whose value for an entity instance can be computed from the values of its other attributes using a pre-defined function. For example, in Figure 1, the attribute *num_phones* captures the number of phones of a person instance but it can be computed given the value of the multi-valued attribute *phone* for that instance. While the domain or type of the derived attribute is no different from the domains of ordinary attributes, e.g., date, character string, number, an active attribute evaluates to a finite data (e.g., audio, video) stream that is not query-able by content.

The following defines these constructs.

- An *active* attribute or entity has a name beginning with the character '@'.
- The value of an *active attribute* of an instance of an entity is nominally a character string s. However, when the attribute is evaluated (for querying), s is interpreted either as executable (a program or a data associated with a

Fig. 1. Derived attributes

particular application a URL). The value returned for the attribute is the result of executing open(s,a) where a is a list of all the non-active attribute values for that instance.

- An *active entity* is an entity with one or more active attributes.
- No entity can have *only* active attributes. In particular, the identifying or partial key attributes must not be active attributes.

One consequence of this extension is that streams can now be treated as entity instances that can be queried by *context*, i.e., via relationships with ordinary entity instances.

2.2 Our Framework

To sketch our framework, we turn to Figure 2. Once an important surprise E_s occurs in the miniworld described by a database D, the event(s) are analyzed to extract clues c. How can those clues be used to obtain an explanation of the contributing events and learn from them? We have argued in the introduction that we need to find relevant sources of available data streams and invent a process X to use the clues c to obtain stream sequence delimiters λ using which clips σ are extracted from the streams. These clips are analyzed in a possibly semi-automated process A for more clues c'. This process continues until the data set obtained is adequate for a reconstruction of the causal events.

In order to formalize this process X, we make use of the extension of the ER model with active constructs. Figure 3 indicates that the process X is facilitated by the addition of relationships between the enterprise database model D and

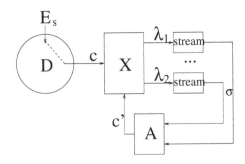

Fig. 2. Querying streams

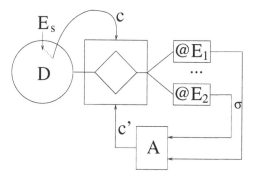

Fig. 3. Accommodating streams

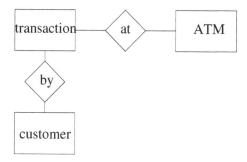

Fig. 4. Fragment of conventional model for an ATM

active entities such as E_1, E_2. These relationships make explicit what contextual clues can be used to query relevant clips from stream data starting with the data in D. In a sense, this allows the enterprise to be ready for handling surprises efficiently and avoid future repetitions by monitoring those streams.

For example, consider Figure 4, a fragment of an ER model for a traditional database that keeps track of ATM transactions. Customers enact banking transactions at automatic teller machines. The surprise event occurs when a customer is physically attacked just after withdrawing money from the machine. To analyze this event, we resort to video cameras at the ATM location and extract a clip of the criminal using the ATM transaction as the approximate time of occurrence. A subsequent clip then shows a vehicle used by the criminal to exit the scene. Analysis can then be used to predict possible street intersections at close proximity containing video cameras that this vehicle could approach. Those video cameras are then interrogated and clips are extracted for examination. This process continues until a getaway path is established.

Learning from this process, and to avoid future surprises of this kind, we arrive at the model of Figure 5. Here the active entities *@person* and *@vehicle* represent the fact that clips containing approaching persons and their vehicles can be obtained from a video file. Note that the active entities *@person* and

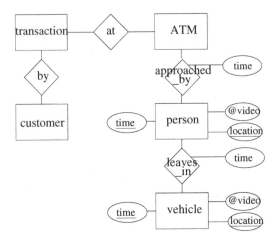

Fig. 5. Above model extended with active constructs

@vehicle have location as attribute. Thus, they capture video streams at any location in the city or elsewhere. Clearly, this will facilitate the tracking of future surprises of this genre.

3   Example

While the problem of learning from surprises typically conjures up scenarios that are related to security, we present here one that is quite different: information technology support for the classroom. Our intent is to highlight the generality of our approach. The surprise in the otherwise predictable activity in the classroom is in the interruptions: questions, answers, demonstrations, and discussions. It turns out that our information system must store stream video/audio data, be able to play it back based on these unexpected events, and also relate such clips with each other.

3.1   The Classroom

A great deal of technology has augmented the classroom: the blackboard has turned white, the grinding noise of chalk has been replaced by silent markers, projectors display lecture 'slides' on computer screens, video and audio recordings of the lecture can be made and stored with modest resources. However, a clear rationale for the integration of technology with instruction is not always apparent to educators [4]. The aim of this application therefore is to use appropriate technology. We envision a classroom that is augmented with reasonable modern technology which can be exploited by the system being designed in order to facilitate questions, answers, discussion, and demonstration in the classroom. The resources / equipment we require are the following:

1. one slide file prepared by the instructor that will serve as the baseline visual aid for the lectures; the file should be of a known format, e.g., Powerpoint, PDF;
2. zero or more demonstration files that the instructor will have the option of using in order to demonstrate some phenomenon. For example, it could be a video file displaying the effect of surface tension or behavior of colliding or falling bodies under various environmental constraints; alternatively, it could be a computer program such as an equation solver or algorithm animator. Of course, the instructor has the option of using physical objects instead, or of skipping demonstrations altogether;
3. a PC that runs our system as well as the applications that display the slide file and the demo file, if any. A slide file should not only be executable but it should be able to display any particular slide given the slide number in the file. We assume that the system can also interact with the instructor via special button-presses on a hand-held remote; and
4. an audio/video recorder because it helps those who have missed a class to catch up.

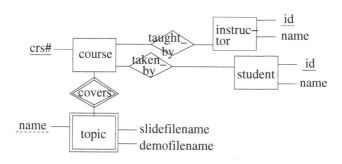

Fig. 6. A fragment of an ER diagram for the classroom

The ER diagram (relationship cardinalities omitted) in Figure 6 expresses a fragment of a standard model that could well be a starting point for us in designing the database infrastructure. A course is uniquely identified by an attribute *crs#*; it is taught by an instructor and taken by students. *Topic* is a weak entity with *name* a partial key (it uniquely identifies a topic given a particular course).

The first problem is with the slide file and the demo files. They are shown here as attributes of *topic* and that is inadequate because the content of the slide file is a sequence of images for which querying by content is not straightforward. But the bigger problem is the surprise, which we deal with below.

3.2   The Surprise

The surprise is that the unidirectional flow of information from the instructor to students — an instructor lectures while occasionally writing on a blackboard as students listen attentively — is rarely adequate. A greater amount of material is absorbed by a larger body of students when the latter are asked to become active learners. Active learning[2] is based on the belief that "to learn something well, it helps to hear it, see it, ask questions about it, and discuss it with others" [5]. From the perspective of the student, these activities are extremely useful as one student's question, comment, or reaction and its resolution enhances clarity of the subject matter for another student. Interestingly, those same activities helps the instructor whose perspective is quite different from that of the student. Unexpected questions, answers, and discussions illuminate lack of clarity and ambiguities in statements, missing steps in explanations, and pitfalls in a student's attempt at comprehension that even experienced teachers would be hard put to predict the first time he/she lectures on a topic. No wonder that through the very act of teaching, an instructor learns more about the intricacies of the subject matter and about how best to unravel them while lecturing, typically performing much better when teaching the same subject matter a second time. Unexpected questions and their answers therefore are crucial.

The analysis of such surprise is consequently important. The instructor may wish to know which topic provoked the most questions; reviewing those questions may lead him/her to remove certain slides for the same course in the following term. However, any attempt to delete a slide in the current term should be prevented if any question or answer referred to it. Similarly, a student who, while reviewing the lecture slides, wishes to pursue any or all of the questions that arose from a particular slide, should be able to do so. But, the question-answer component is recorded as a video/audio stream that is simply not queryable by content. Furthermore, it may be useful to pursue the follow-up questions, and their corresponding answers; consequently, we need to look at the structure of these interruptions.

At any point in time when the instructor lectures (Lec1 in Figure 7) using electronic slides from a pre-created file, a student can ask a question (Q1 in Figure 7). Typically, the instructor would answer the question verbally (A1), perhaps aided by further writing on the board or by referring to another prepared slide (not necessarily one shown earlier). An answer may also be an acknowledgment of an error in a slide or text. The answer itself could be interrupted by a question, e.g., requests for clarification or disambiguation (Q2 in Figure 7) requiring a response (A2); the instructor then resumes the interrupted answer (A1). It is also possible that after this question, there will be a follow-up question (Q3) typically related to the earlier question. Later, the instructor decides that the answer is enough and resumes lecturing (Lec2). Sometimes, it is the instructor who poses a question relevant to the current slide and the students attempt to answer. In this case, a single question can be followed by several student answers (some of them punctuated by the instructor requests for clarification) followed by the instructor's own answer.

Thus, the cause-effect nature of questions and answers is very important for queries but this can only be captured through a relationship between audio/video

---

[2] The use of the term *active* in the context of learning is purely accidental. It has nothing to do with active entities and attributes.

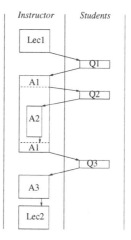

Fig. 7. An activity diagram with two swimlanes for the question-answer workflow

streams. Time itself is not enough to identify the relationship since a question about a slide can be asked when the instructor is covering a later slide. Also, we must keep track of the nested nature of questions and answers.

One approach is to represent questions and answers as entities and capture the cause-effect as a relationship but the inability to query these streams by content makes that awkward. Another approach is to treat questions and answers as events but the nesting structure makes that a problem. Our solution is to accept the first option but treat the content as non-query-able but executable; i.e., exploit our extension of the entity-relationship modeling constructs.

3.3   Tackling the Surprise

First, we analyze the contextual clues linking the audio/video stream with our infrastructure. Here, the start and end times of each question and answer are the links. Consequently, we can offer an active ER diagram Figure 8 that extends the previous model in Figure 6.

The entities *instructor* and *student* are now generalized into the entity *participant* using an *is_a* construct (this is therefore an ER diagram extended with generalization hierarchies, but this aspect is not important for our discussion). The entity *topic* is now related to two active entities *@demo* and *@slide*, both of which are related to the active entity *@question*. The relationship *explained_by* represents the cause-effect between questions and answers. A relationship *leads_to* between *@answer* and *@question* allows questions to interrupt answers whereas the recursive (ring) *followup* relationship on *@question* allows questions to be serialized and yet associated together. Finally, a *referred_to* relationship records which slides were used while answering a question.

An instance of the active entity *@slide* will have the slide file name as *sysname* and opening that file with *number* as parameter will display exactly one

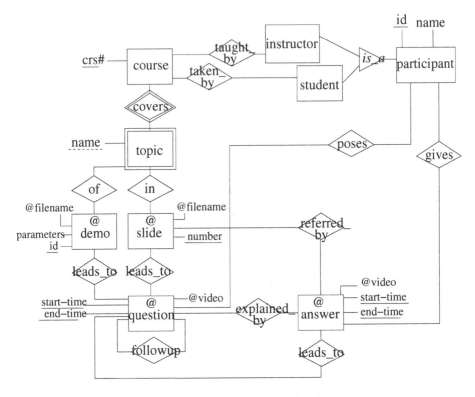

Fig. 8. An active ER diagram of the classroom

slide. (The system can easily enable the instructor to go to the *next* and *previous* slide and not merely retrieve by number.) Similarly, the *@question* active entity will have an audio/video file as *sysname* and when that file is opened with *start-time* and *end-time* as parameters, a clip will be returned corresponding to that interval. (Adding a preamble and sequel is also possible.) For the *@demo* active entity, *sysname* will be the demo file which when opened with parameters obtained from *parameters*, a list of (variable number of) parameters, the demonstration will occur.

Since there is a recording (audio only or audio plus video) of the entire lecture, we can use its time track or counter for our system to monitor and read; we need the ability to extract the counter / time reading at all times when recording is in progress and the ability to play back a clip given a starting time and an end time. We can require the instructor to press a button when a question is asked, press another button when the question is over, and a similar pair of buttons for the answer. Similarly, the instructor may initiate a demonstration after pressing a *demo* button regardless whether the demonstration uses physical objects or executes a file. This demonstration may itself initiate questions and answers.

The button presses can be used to indicate to the system the start-time and end-time attribute values for the clips. The instructor can choose to either nest or serialize the next question. Consequently, the sequence of questions will follow a nested structure. Thus, if the start and end button presses (for, say, question) are represented by '(' and ')', the button-presses will belong to the language of well-balanced parentheses. Thus, the system will display the current state of nesting in these question-and-answer sessions to help the instructor in choosing whether to nest or serialize. The relationship between a question and the slide that provoked it is easily inferred and stored. However, the answer may refer to a number of slides: all of these are stored corresponding to the relationship *referred_by*. Thus, for a question-answer session, the slides involved, the corresponding segment of the audio/video stream, whether the participant is a student or the instructor can all be stored. Thus all the goals of flexible querying can now be met.

Thus, it will be possible to select questions related to a particular topic (a relational join in the standard database implementation of an ER model [6]) based on the standard attributes of the entities and relationships involved, or even the questions asked by a particular student. Referential integrity constraints will disable any attempt to delete a slide that is used in an answer or question. More generally, any semi-structured data [7] can be treated as instances of an active entity enabling comparison (and joins) using the structured attributes.

From the audio/video of one question, the user can ask for its answer or instead go to follow-up questions and their answers, or move to nested questions and perhaps their answers.

4   Related Work

While our use of the classroom is merely an example, it is different from the typical work on conceptual modeling for e-learning and classroom learning as the latter revolve around modeling the domain (e.g., [8]) While in [9], an active conceptual model involves operations as well as data, in our case, it is not enough to focus on operations or even events: it is important to confront the unstructured nature of certain data. In [10], the review of a thesis is stored and made available not only for the author of the thesis but to future thesis writers, while finessing the privacy considerations. However, this work focuses on making the shared knowledge very structured avoiding unstructured data which is the problem we face. In [11], active XML documents are created that lookup other Web documents to find element contents. This is very similar to our active attributes, but they examine the efficacy of lazy evaluation while we focus on active entities and their joins with other entities through relationships.

5   Conclusion

In this paper, we have argued that an essential ingredient of an information system infrastructure for active conceptual modeling of learning from surprises is the ability to query data that is not query-able by content, especially streams. There are two principal contributions of this paper. First, data streams that are not query-able by content can be modeled using an extension of the ER model.

Second, for an enterprise to manage and learn from surprises, its traditional conceptual model should be made ready to *link* with these streams by incorporating relationships with such active entities. These relationships provide the contextual clues for querying these streams.

We have outlined an application where the surprise involves a criminal assault on a customer at an automatic teller machine. We provide a more detailed example of a classroom environment where the surprise comes from interruptions: questions, answers, discussions, and demonstrations. In both examples, the standard database provides the context for querying stream data through relationships with active entities modeling useful data streams that are not query-able by content.

In general, our approach of dealing with surprises would necessitate harnessing and exploiting the abundant stream data. Such data are owned not only by city, state, and national governments, but increasingly by individuals equipped with cell phones and other portable electronic devices of growing sophistication connected to a networked world. The challenge therefore the extraction of useful segments from this extremely large collection. This can be tackled on three fronts: discovery of the right content, incentives for the diverse collection of owners to share, and markers that provide reliable linking using contextual clues. This will be the subject of our future work.

Acknowledgment

The author is grateful for comments from anonymous reviewers on the original version of the paper and for a subsequent reference from Dr. Leah Wong.

References

1. Chen, P., Thalheim, B., Wong, L.: Future direction of conceptual modeling. In: Chen, P.P., Akoka, J., Kangassalu, H., Thalheim, B. (eds.) Conceptual Modeling. LNCS, vol. 1565, pp. 287–301. Springer, Heidelberg (1999)
2. Chen, P., Wong, L.: A proposed preliminary framework for conceptual modeling of learning from surprises. In: ICAI 2005. Proceedings of the 2005 International Conference on Artificial Intelligence, pp. 905–910 (2005)
3. Lesk, M.: How much information is there in the world? Technical report, lesk.com (1997), http://www.lesk.com/mlesk/ksg97/ksg.html
4. Fairey, C., Lee, J., Bennett, C.: Technology and social studies: A conceptual model for integration. Journal of Social Studies Research (Winter 2000)
5. Silberman, M.: Active Learning 101: Strategies to Teach Any Subject. Allyn & Bacon, Boston (1996)
6. Batini, C., Ceri, S., Navathe, S.: Conceptual Database Design. Benjamin/Cummings (1992)
7. Abiteboul, S., Buneman, P., Suciu, D.: Data on the Web: From Relations to Semistructured Data and XML. Morgan Kaufmann, San Francisco (2000)
8. White, B.Y.: Thinkertools: Causal models, conceptual change, and science education. Cognition and Instruction 10(1), 1–100 (1993)

9. Moulton, A., Bressan, S., Madnick, S., Siegel, M.: An active conceptual model for fixed income securities analysis for multiple financial institutions. In: Ling, T.-W., Ram, S., Lee, M.L. (eds.) ER 1998. LNCS, vol. 1507, pp. 407–420. Springer, Heidelberg (1998)
10. Rickert, W.F.: Techniques for knowledge sharing in thesis writing. In: Proceedings I-KNOW 2005, pp. 657–663 (2005)
11. Abiteboul, S., Benjelloun, O., Cautis, B., Manolescu, I., Milo, T., Preda, N.: Lazy query evaluation for active XML. In: Proceedings SIGMOD 2004, pp. 227–238 (2004)

Approaches to the Active Conceptual Modelling of Learning

Hannu Kangassalo

University of Tampere, Department of Computer Sciences
FIN-33014 University of Tampere, Finland
hk@cs.uta.fi

Abstract. Information modelling is a collection of dynamic processes, in which its content develops from physical processes to abstract knowledge structures. We study that collection on several levels of abstraction of human cognition and knowledge. These processes can be performed through various approaches, on several levels, and by using several perspectives. We concentrate on active conceptual modelling, which has become important in the science and technology, including educational sciences and learning. It is a process of recognition, finding or creating relevant concepts and conceptual models which describe the UoD, representing the conceptual content of information to be contained in the IS. This characterisation contains the construction of new concepts, too. We study methods for collecting information from various sources in the UoD and accumulating it as possibly actual instances of various types of pre-defined concepts. Some of these instances may be cases of sudden events or processes. They should be recognised as concepts and included in to the conceptual schema. To some extent, some concepts may be constructed which fit to this collected information. During the adaptation process we are applying active conceptual modelling for learning, which organises our conceptual schema in a new way. Learning is a process in which a learner re-organises, removes or refills his knowledge structures by applying his newly organised conceptual schema.

1 Introduction

Information modelling is used in many branches of research and practical work, as in the theory of science, in scientific research, conceptual modelling, knowledge discovery, and other areas. Information modelling means structuring originally unstructured or ill-structured information by applying various types of abstraction models and principles developed for different purposes. In this work we regard it as a collection of dynamic processes, in which its content develops from the level of physical processes, through several levels of abstraction, to the level of most abstract knowledge structures in which its information content is the most useful, usable and valuable for the human.

Conceptual modelling is the most important sub-area of information modelling. It allows us to construct structures made of information, called *concepts*, which are based on conceptualisation and concept formation processes, system structuring, and

P.P. Chen and L.Y. Wong (Eds.): ACM-L 2006, LNCS 4512, pp. 168–193, 2007.
© Springer-Verlag Berlin Heidelberg 2007

justification for system design. In these activities information is formulated and organised on the basis of selected purposes. Concepts facilitate the understanding, explanation, prediction and reasoning of information, its meaningful manipulation in systems, and understanding and designing of functions of various types of systems. It also helps the development of methods and theories for investigations, scientific research and design tasks.

We regard conceptual modelling as a process of:

1. Creating, recognising, or finding the relevant concepts and conceptual models which describe the Universe of Discourse (UoD), e.g. of the information system, or the design of functions or relationships in the UoD, and
2. Selecting and representing the relevant conceptual content of information to be contained in the system of information units, and based on this UoD.

The system of information units may be a traditional information system, but also some other type of system, e.g. a model of a geographical multi-level map system, a biological multi-level description of various life forms in a tropical jungle, or a sociological description of a human society in a metropolitan area.

Often the conceptual schema [1] can itself be regarded as a desired system, too, as it describes conceptually that part of the world from which we are interested in. The conceptual content of the IS may include deductive information, too, i.e. derived concepts and derived inference rules, although they may also be in an implicit form.

The third important notion in this work is language, which is necessary for communication and also needed for the construction of linguistic structures, e.g. sentences.

We will study how the abstraction level structure used in modelling can be organised and used. We analyse the general abstraction level structure, the structure of information in semiotics, and abstraction levels of human cognition, all based on the theory of hierarchical, multilevel stratified systems [2,3]. Concept formation and conceptual modelling are considered from the point of view of members of the user society, and from the point of view of the professional modeller. Some aspects of concepts, definitions of concepts, and concept structures for conceptual modelling are briefly described. We will study some methods for collecting information from various sources in the UoD and accumulating it as possibly actual instances of various types of pre-defined concepts. In some situations, these instances may be cases of sudden events or processes, for which concepts describing them have not yet been developed. In some cases, information may be constructed as concepts, which fit this collected instance information. We have to construct them as concepts that should be adapted into our existing conceptual schema and knowledge base. During the adaptation process we may construct new concepts and modify or remove old concepts and rules between concepts, and some knowledge structures, i.e. we are applying active conceptual modelling for learning. Active conceptual modelling organises our conceptual schema in a new way. Learning is a process in which a learner extends, re-organises, removes or refills his knowledge structures by applying his newly organised conceptual schema.

2 The Theory of Stratified Systems and Levels of Human Cognition

We regard information modelling as a dynamic collection of processes, in which the information content of received input from the external world develops from physical processes to abstract knowledge structures. There are several works in the literature, which develop or apply a similar approach, perhaps using slightly different terms [2,3,4,5,6,7]. We study that collection first as a general system of several levels of abstraction, then as a structure of human cognition, information, knowledge and language, and finally as a structure of 15 levels of abstraction of human cognition. In all cases we use principles of absolute abstraction levels based on connected levels of abstraction, as a thinking tool [7].

Mesarovic, Macko and Takahara describe the general characteristics of a stratified description of a system [2]:

1) The selection of the strata, in terms of which a given system is described, depends upon the observer, his knowledge and interest in the operation of the system, although for many systems there are some strata that appear as natural or inherent.

2) Contexts in which the operation of a system on different strata is described are not, in general, mutually related; the principles or laws used to characterize the system on any stratum cannot generally be derived from the principles used on other strata.

3) There exists an asymmetrical interdependence between the functioning of a system on different strata.

4) Each stratum has its own set of terms, concepts, and principles.

5) The understanding of a system increases by crossing the strata: in moving down the hierarchy, one obtains a more detailed explanation, while in moving up the hierarchy, one obtains a deeper understanding of its significance.

The most important characteristics of the stratified description are 2, 3, and 4. On the basis of these, at least two different hierarchical, multilevel structures can be constructed.

In "The Meaning of Information", D. Nauta, Jr. independently describes the place of information in semiotics [8], which follows quite closely the principles of a stratified description of a system. Nauta Jr. describes in the same diagram both the organisation of knowledge that gives a kind of participational view of the organisation of behaviour, and the theory of knowledge that gives a kind of observational view of the theory of behaviour. This diagram shows that information appears in different forms in different levels of abstraction. Later, The FRISCO Task Group of IFIP TC8 WG8.1 Design and Evaluation of Information Systems, added one level on the top of the structure in order to represent aspects of the social world and replaced three lower levels [9].

In Fig. 1, from "The Meaning of Information", "the ontological levels ranging from primitive organization of behaviour to explicit organization of knowledge, are listed at left in the diagram. The latter realizes a definite subject-object relation, transforming the participational view into an observational 'meta-view'. The levels at right are epistemological in character; they range from the theory of simple behavior (which is an application of the 'observational' organization of knowledge to the study of primitive organizations of behavior) to the theory of science (which is an application of the organization of knowledge to the study of its own highest developments)" [10].

Nauta, Jr. describes seven levels and types of information. The level of potential information is the level of the physical world. On this level there are signals, traces, physical distinctions, physical tokens, speeds, laws of nature, etc. On this level it is possible to study physical message exchange by using e.g. the mathematical theory of communication. Signal semiotics is the study of implicit information involved in the signal processes occurring in a real organism or machine, etc, [10] Sign semiotics studies concursive information, which consists of combinations of pieces of information. Symbol semiotics is an activity proceeding discursively by reasoning or argument, but not intuitively.

Grammar occupies a central place in that it functions as a kind of transformator transducing the organization of behavior into the organization of knowledge; it is also the site of the transition from *de facto* redundancies to *de jure* redundancies, as well as from semiosis to transcendental – or metasemiosis. Accordingly, the domain of metasemiotic information may be characterized as post-linguistic, and that of semiotic and zero-semiotic information as pre-linguistic. [10]

The last structuring principle in this work and used for information modelling is a cognitive level structure in human cognition. It is important, because e.g. human cognition is essential in recognising sudden and fuzzy phenomena and events. We only describe it briefly and analyse it as far that we can recognise the structuring principle, which seems to be the same principle of absolute abstraction levels based on connected levels of abstraction.

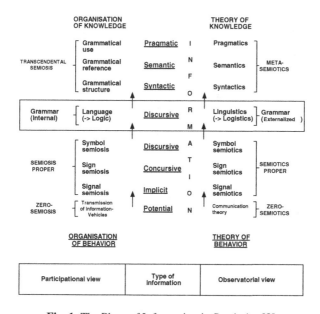

Fig. 1. The Place of Information in Semiotics [8]

Cognition is an extensive notion that concerns or involves knowledge [11]. It covers different aspects of knowing and consciousness. J. H. Flawel says "The traditional image of cognition tends to restrict it to the fancier, more unequivocally "intelligent" processes and products of the human mind. This image includes such higher-mental-processes of psychological entities as knowledge, consciousness, intelligence, thinking, imagining, creating, generating plans and strategies, reasoning, inferring, problem solving, conceptualizing, classifying and relating, symbolizing, and perhaps fantasizing and dreaming" [12].

We do not describe here connections between different aspects and sub-areas of cognition. There are many of these and together with the things described here they form a system of cognition, which consists of many phenomena on different levels of organisation. In the same way, we study information and knowledge in their many different forms, which are placed on different levels.

We start our study on low-level structures and continue to structures on higher and higher levels of abstraction. We will not take into account all the details of different alternatives.

1. Non-conceptualised observation data: It is data that contains a direct observation from the UoD, without any conceptualisation. In some cases the situation can be more complicated. The essential idea behind the theory of a nonconceptualised mental content is that some mental states can represent the world even though the bearer of those mental states does not possess the concepts required to specify their content [13].

2. Tacit knowledge [14, 15] Tacit knowledge consists of complex, diffuse and mostly unrefined knowledge accumulated as know-how and understanding in the minds of knowledgeable people [16]. Tacit knowledge may appear on several levels of the organisation of cognition. Here we introduce it on the lowest possible level. The amount of tacit knowledge seems to depend on the amount of diffuse information from the UoD, on the amount of more clearly structured knowledge, and on the level of abstraction of knowledge.

3. Conceptual knowledge: A concept is a possibly named, independently identifiable structured construct composed of knowledge primitives and/or other concepts [17]. A concept can also be regarded to be an intensional knowledge structure that encodes the implicit rules constraining the structure of a piece of reality (cf. [18, 19]). Concepts give the categories used for modelling the UoD. There are several types of concepts: general concepts or universals, UoD-specific concepts, personal concepts, structures made up of concepts and relationships between them, and ontologies used to formalise and organise basic knowledge.

4. Perceptions of recognised instances: Perceptions of recognised single instances are observations of instances classified on the basis of one or more concepts, or on the basis of the whole system of concepts, called ontology.

5. Facts: Recognised states of affairs in the UoD. Note, that there are states of affairs based on concepts, only, or states of affairs based on concepts and observations, or states of affairs based on observations, only.

6. Symbolised instances: Symbolised instances are instances of objects referenced by using symbols, which usually are interpreted as names.

7. <u>Rules</u>: Rules either guide or constrain behaviour or thought [20]. Rules are normative constraints in types of decision or, more generally, judgement; decisions can be taken to be judgements as to what is best. Rule-following is an intentional activity of the sort that may be involved in using words, moving chess-pieces, adopting local custom and thinking straight [21].

8. <u>Inferencing mechanism</u>: Inference can be understood as the upgrading or adjustment of belief in the light of the play of new information upon current beliefs, it is customary to recognise at least three modes of inference: deductive, inductive, and abductive, although abduction is often treated as a special case of induction [22].

9. <u>Skill knowledge</u>: The knowledge needed for applying a skill. For example, the skill of shooting requires knowledge about aiming, about behaviour of a weapon, about the wind, etc. We assume that skill knowledge is complete, i.e. a user is an expert in applying the skill.

10. <u>Art, skill and technique</u>: Usually they are practical abilities based on training and learning. We assume that the user is an expert in applying the skill. If he is not yet an expert, then skill knowledge may be on a higher level than skill itself, because the learner has to develop and exercise his skill knowledge and his ability to use it.

11. <u>Cognitive model / conceptual schema</u>: Structure, which consists of concepts, relationships between concepts, facts, and knowledge primitives. Cognitive models are mental constructs; conceptual schemata are either mental constructs or externalised constructs.

12. <u>Theories</u>: A (scientific) theory is an attempt to bind together in a systematic fashion the knowledge that one has of some particular (real or hypothetical) aspect of the world of experience [23]. The notion of a theory can be understood in two different ways: by using a traditional statement view (a theory is a consistent set of statements), or by using a non-statement view (a theory is a mathematical structure), together with the specification of its intended applications [17,24,25]. There exist many types of theories.

13. <u>Norms</u>: Norms are rules that support uniform social behaviour, social action, and talk.

14. <u>Values</u>: Values are (here) abstract valuable things that the actor regards as important to herself or to her colleagues or partners, according to which activities are directed and evaluated.

15. <u>Languages</u>: A language is a means of communication, by use of which a sender sends messages to one or more recipients. A message contains some conceptual content, which has some meaning to both the sender and the receiver. Depending on the situation, the conceptual content of a message may be the same or different to the sender and the receiver.

In addition, ideals, religious beliefs, etc. might be taken into account. All these 15 types of information are in use when people act in organisations. The most important alternatives are concepts and languages. So far information systems are not able to use them completely – even the proper concepts are used often in a very inconclusive way. Therefore we have to take into account both human knowledge and computerized data, and compare and apply them both.

3 Foundations of Conceptual Modelling

There are many theories about the nature and origin of concepts, e.g. see [26,27,28,29] Concepts are central notions for conceptual modelling and scientific research, which are closely related processes. The fundamental notion, *concept*, is here defined to be an intensional knowledge structure that encodes the implicit rules constraining the structure of a piece of reality (cf.   [18,19]). It is a central epistemological unit of knowledge. It has two roles at the same time:

1. It composes and organises information regarded as necessary or useful for structuring and understanding some piece of knowledge, and
2. It characterises some features of objects that fall under it.

A *basic concept* is a concept that cannot be analysed using other concepts of the same conceptual system. In reality, concepts are not classified e.g. into objects, entities, attributes, relationships, or events, etc. This kind of classification is not an intrinsic feature of knowledge - it is a superimposed, abstract, ontological or epistemological scheme into which systems of concepts and knowledge in them are forced, especially in conceptual modelling. Concepts are just constructs containing information forming together conceptual systems.

A *derived concept* is a concept (definiendum) the characteristics of which have been derived from the characteristics of other (defining) concepts in the way described in the definition of that concept. The defining concepts are in turn defined by other, lower level concepts until the level of undefined, basic concepts is reached [17]. The result of conceptual modelling, i.e. *conceptual content* of concepts and conceptual models, depends on:

–   information available about the UoD,
–   information about the UoD, regarded as not relevant for the concept or conceptual model at hands, and therefore abandoned or renounced,
–   philosophical background to be applied in the modelling work, e.g. realism,
–   additional knowledge included by the modeller, e.g. some knowledge primitives, some conceptual 'components', selected logical or mathematical presuppositions, mathematical structures into which information is forced, etc.,
–   knowledge about collection of problems that may be investigated in this environment,
–   selected ontology used as a basis of the conceptualisation process,
–   epistemological [30] theory, which determines or directs how ontology is or should be applied in the process of recognising and formulating adequate concepts, conceptual models or theories, and constructing information, data, and knowledge, on different levels of abstraction,
–   the purpose and goals of the conceptual modelling work at hands,
–   collection of working methods, and their theoretical backgrounds, for conceptual modelling,
–   the process of the practical concept formation and modelling work,
–   the knowledge and skills of the person making modelling, as well as those of the people giving information for the modelling work.

Conceptualisation, i.e. human concept formation, can be seen as a process in which individual's information about something is formed and gets organised into concepts and corresponding knowledge (see [31]). This view is much applied in education, but it is not complete and precise. On the basis of the characterisation of the notion of a concept, the results of the conceptualisation process can be regarded as an intensional knowledge structure which encodes the implicit rules constraining the structure of a piece of reality (cf. [18,19]). These results are concepts.

The content of a concept is its *intension*. It consists of knowledge primitives, other concepts, and the structure they form in the concept. These components are called its *characteristics*. It may contain also factual knowledge, as in the concept of an extended family, which includes at least three children. The number of children is factual knowledge that has been decided by the parliament of the country.

The set of objects, as well as data representing these objects, to which the concept applies, is called its *extension*. A concept always has the intension, but its extension in the UoD may temporarily or always be empty. We have also to note, that intensions determine extensions, but not conversely [32]. If we don't know the intension, we cannot be sure what are the characteristics (properties) of the concept and we don't know what the concept is. The elements of the extension are called *instances* or *occurrences* of that concept.

The human conceptual scheme (including intension) is important for determining what we are able to recognise in the UoD or in our own thinking processes, because it contains the results of all conceptualisation processes made by the person, or – at least quite many of them. If a person does not possess a certain concept, then he is not able to properly recognise the instance of that concept, before the concept has developed. This aspect is important for learning and also for the study of learning.

By means of the *external conceptual schema* the conceptualisations of relevant concepts and the rules between concepts in the human conceptual schema are made 'visible'. With it the recognition, structure, behaviour and functions of the object system, and the information content based on this object system are externally described and can be shared and learned.

In general philosophy, *ontology* is a theory of existence; what exists. In formal ontology and conceptual modelling, ontology can be regarded as a logical theory that gives an explicit, partial account of the conceptualisation. Guarino and Giaretta also say that ontology can be regarded as a synonym of conceptualisation [18].

Epistemology is a branch of research that studies information and knowledge. It is the study of nature of information, knowledge and justification: specifically, the study of (a) the defining features, (b) the substantive conditions, and (c) the limits of knowledge and justification [33]. It attempts to answer the basic question as to what distinguishes true or adequate knowledge from false or inadequate information? This question translates into issues of scientific methodology and knowledge discovery; how one can develop explanations, theories or models that are better than competing explanations, theories or models. That question is exactly the same as the basic question for selecting the best conceptual content for the information system.

4 Modelling Situations and Modelling Processes

4.1 Modelling: The Beginning

In the initial stage of modelling, concept formation and conceptual modelling are cognitive actions of a human being. The work starts from observations concerning some details of the UoD. A group of users may recognise an epistemic or conceptual difficulty, which they try to handle by using concepts and knowledge at their possession so far. If they cannot solve that difficulty, some change of knowledge, or some conceptual changes may be needed. The system of concepts by using which people are modelling the UoD, has to be changed and structured in a better way, which somehow fits the situation better.

We assume here that a professional modeller is invited to handle the situation. The modeller is compelled to recognise, analyse, and design, with the users, concepts of an application area, the content and structure of concepts, and how these concepts are applied in users' thinking and language, and finally, to construct an external conceptual schema of the UoD.

A person is paying attention to a phenomenon in the UoD, possibly first only as tacit knowledge or as a result of an observation, and constructing concepts and a conceptual schema, see Fig. 2. Tacit knowledge consists of elemental basics of information, which cannot as yet be recognised as mental images or even simple concepts. An observation is first conceived as a mental image in the human mind. Later on, a concept may be constructed in which information in the mental image may be included when more information has been accumulated in different situations. There are several theories about concept formation in cognitive science and psychology (see e.g. [34,35]).

According to constructivist personal construct psychology [36], a concept structure in a person develops on the basis of his experiences, including his own observations, communication with other, his own reasoning, reading, organised learning, information gathering by methodological or technical external devices, etc. Because even in the same situation, different people pay attention to different aspects of the situation, they will end up with at least slightly different observations about any situation.

The external representation of a conceptual schema is constructed by using more or less formal notation. The external representation of a concept is used e.g. for communication and to support thinking. Its user must master the information content of the concept and the concept system in which the concept is included before he can

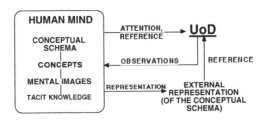

Fig. 2. A phenomenon in the UoD, concepts, a conceptual schema, and an external representation

meaningfully use the *symbol* of the concept. On the other hand, the 'reader' of the symbol must have enough information about the situation to grasp the concept, i.e. the *meaning* of the symbol. The symbol belongs to the linguistic level (level 15) in the cognitive level structure in human cognition discussed in chapter 2, and its content (meaning), i.e. concept or other constructs, belong to the levels 3 – 14.

A human concept is always a subjective thing. Similarly, a conceptual schema of the UoD of a person is a subjective construct, which reflects the personal perspective of that person who has created it. However, people live in societies and therefore need common concepts and conceptual schemata about their environments. This demand (among other things) triggers the processes of learning and sharing of concepts and conceptual schemata with other people.

The users utilise the concepts developed to be most suitable for their own views. Often the modeller's and the users' views on the UoD, or even on the same concept, may differ quite a lot. The only way out of this difficulty is to analyse the intension of both concepts in detail. In this analysis the concepts regarded as important must be studied until the level of undefined, basic concepts is reached. The content of both concepts must be developed so much that both persons can agree that they have (at least almost) the same concept. That activity is a part of learning.

4.2 Active Conceptual Modelling

Active conceptual modelling is quite similar to a traditional conceptual modelling, but in it some sub-processes will be made slightly differently, and with different pre-conditions. In it time is a more central notion than in traditional conceptual modelling. We may need new concept definitions almost immediately after the observation, adopted in the best possible way to the existing or modified conceptual schema. The system for supporting active conceptual modelling needs functions for creating, recognition, or finding new relevant and timely concepts, and re-organising or changing the existing conceptual model so that new concepts can be applied immediately. Often a modified conceptual schema is constructed and used before the language for representing the new conceptual schema is adapted to the new situation.

P. Thagard has presented degrees of conceptual change, roughly ordered in terms of degree of increasing severity [37]. The first two items can be fully understood as belief revision (here, B-addition, B-deletion), but other parts involve conceptual changes whose epistemic import goes far beyond belief revision, see Fig. 3.

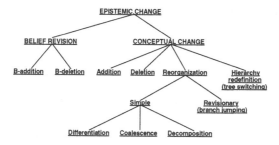

Fig. 3. Taxonomy of kinds of epistemic change

A *conceptual change* means a change in which the conceptual world of one or more actor is changing. If the change concerns only one person or a few people, it may seem rather inefficient, but that is not always the case because almost all major discoveries have been innovations of one or a few people, only. Adding new concepts and rules often causes modifications in several places in the conceptual model. Described in more details, a conceptual change means a change in which: 1) a new concept is constructed into the conceptual model [38], and perhaps new instances of it are added, or 2) an old concept is deleted and all instances of it are deleted as well, or these instances are placed in extensions of other concepts and possibly modified somehow, or one or more of the following cases appears (cf. [37]):

3) Simple re-organisation:

   3.1) Differentation: division of a concept produces two or more new concepts.
   3.2) Coalescence: a new superordinated concept is added, that links two things previously taken to be distinct.
   3.3) Decomposition: a concept previously regarded as atomary, is divided into two or more parts, but it still remains as one concept.
4) Revisionary (branch jumping): a concept is shifted from one branch of a hierarchical tree to another.
5) Hierarchy redefinition (tree switching): in this case, determining structural principle of a concept hierarchy is changed.

Conceptual change may be caused by reasoning in the recognition of weakly defined or inconsistent definitions of a concept, or by an external requirement that the conceptual model must be changed for some external reason. Sometimes a change must be initiated, when it is recognised that actual properties of instances of some concepts are not valid for the defined concept.

Time is an important constituent of the active conceptual model. In describing dynamic phenomena, we have to construct a time mechanism which allows us to refer to the notion of time, different units of time, different time calculation systems, passing of time, time relations appearing in various phenomena, as precedence and succedence relations, parallelism, etc., and the notions of history and future when using specific notion of time, e.g. Julian or Gregorian time or a relativistic time. Several time systems may be applied in the same conceptual schema.

In concept definitions and in the conceptual model, time appears in several ways, e.g.:

1. Concept definition is relevant only in a certain period, which depends on the system of concepts. We call this notion a *relevance time*.
2. Concept definition may contain conditions and/or constraints, which contain references to time (occurrence values of time) or some calendars. This notion we call a *contained time*.
3. Time can be perceived either as *absolute* or *relative*. Absolute time means understanding of time measured by using the external calendar, independent of the conceptual schema. Relative time means understanding of time measured in relation to other phenomena, defined in the same conceptual schema. *Occurrence time* of instances corresponding to a concept can be defined in relation to a calendar or in relation to relative occurrence times of instances of

the schema. Instances of relative occurrence times must be explained on the basis of the other concepts in the conceptual schema.

We present a small example of an active concept definition, in which a concept of BUYING EVENT is constructed [39]. A line between concepts represents the relationship of conceptual containment [40], e.g. the concept of BUYING EVENT contains the concept of SELLER, and the concept of PAYMENT contains the concept of CUSTOMER and the concept of SELLER. The notion of the PAYMENT describes the activity, in which the CUSTOMER PAYS the COMMODITY to the SELLER by using e.g. certain SUM of MONEY (all details are not shown here).

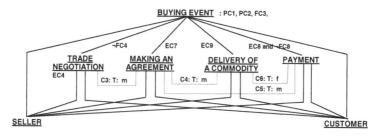

Fig. 4. The concept of BUYING EVENT with two hierarchy levels

Table 1. Some conditions appearing in the concept BUYING EVENT

Pre-conditions:
PC1: EXISTS (SELLER, CUSTOMER)
PC2: SELLER and CUSTOMER have met.
Existence conditions:
EC4: SELLER has the COMMODITY that the CUSTOMER needs.
EC7: SELLER and CUSTOMER have agreed about the PRICE and the CONDITIONS.
        CUSTOMER has made the DECISION.
Finishing conditions:
FC3: PAYMENT has been finished, or TIME FOR PAYING ends, or
        TRADE NEGOTIATION finishes without RESULT.
FC4: AGREEMENT has been made, or the CUSTOMER has finished the
        NEGOTIATIONS, i.e. FC4 = C3 or C4.

4.3   Different Models in Different Situations

Concepts are created by the human mind, partly on the basis of information structures grounded on observations, e.g. in scientific investigations or in coincidental events, partly on the basis of abstract reasoning, and sometimes, partly, on the basis of communications with other people within the society.

Fig. 5 illustrates the situation in which people are sharing at least some of their concepts and make some common external representations. All the community members have some concepts that are 'their own' and some concepts that are shared with other people.

Sharing a concept means that two people possess the token of a concept the intension of which is (at least nearly) similar as the intension of the concept of the other person. Both persons have the token of the shared concept in their own

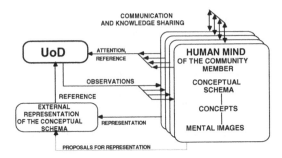

Fig. 5. A person shares some of his concepts within the community [41]

conceptual schemata, i.e. they possess (at least nearly) same knowledge. However, both tokens belong to a different conceptual schema, and therefore the result of reasoning in which these tokens are applied, may mean different things to both people (cf. [27]). Therefore the analysis of tokens of shared concepts should be performed carefully, down to the finest details of the intensions, and also taking the neighbouring concepts into account.

When people represent their private or common concepts as external representations, an additional collection of external concepts and conceptual schemata accumulates. It is called EXTERNAL CONCEPTS in Fig. 6. They may effect the creation and evolution of subsequently developed concepts in several ways, e.g. they may prevent the evolution of newer concepts because users are accustomed on them, or they have learned to use them better than newer concepts, or users try to extend the 'life-time' of earlier useful concepts. Stored external concepts can be studied and applied, modified and refined in making new conceptual schemata.

Fig. 6 reveals that on the basis of one UoD several collections of private or common concepts or systems of concepts may evolve. In addition to collections of human concepts and the collection of external concepts, the external conceptual schema may be represented on paper or on a computer. The human conceptual schemata evolve perhaps continuously, the external conceptual schema must be maintained now and then, the collection of external concepts is developing all the time. Therefore the whole complex of schemata must be maintained with a sense of purpose. This complex of at least five collections of concepts and conceptual schemata will be the basis for active conceptual modelling and learning.

The problem of correspondence between the human conceptual schema and the computer representation of the conceptual schema may give rise to difficult questions. It should be possible to develop an extension facility for the computer representation of the conceptual schema capable of constructing extensions to the conceptual schema on the basis of observation data and concept formation algorithms.

Much attention should be paid to the problem of how users and the modeller(s) can understand and interpret a computer generated extensions of the conceptual schema. The computer representation of the active conceptual schema must have representation mechanisms that make the conceptual schema as readable to the users as the external conceptual schema on 'paper'. However, the computer representation

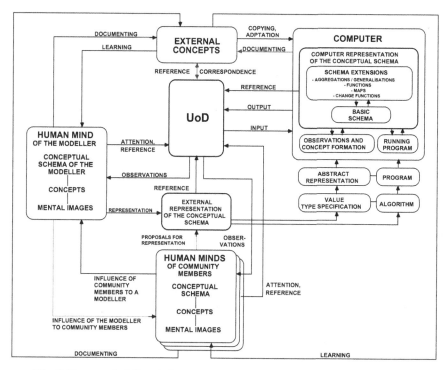

Fig. 6. Human mind, linguistic constructs and a computer as a 'media' of concepts

facility should be able to construct active external conceptual schemata, which have not been constructed before. How can the user understand and interpret the extension, that he sees for the first time. A facility needs advanced metadata functions that can represent all essential metadata, including new types of constructs or phenomena, which the representation mechanism of the existing conceptual schema cannot represent in the computerised conceptual schema, yet. The representation language must be extendable.

Both the modeller and users should pay attention to the applicability of the external or computerised concept descriptions to modelling the current UoD. If the external or computerised concept descriptions have been represented only by using the *surface structure* of the conceptual schema language, then the description of intension of concepts may not be properly understood. The surface structure is a linguistic construct in which only the top-most words of a multilevel semantic are represented. If the concepts are epistemologically abstract, i.e. they are on a high level in the concept definition hierarchy, the users of the concepts must, in principle, also know all the other concepts lower down in the hierarchy, down to the basic level. Otherwise the users cannot interpret the high-level concepts correctly. That happens often in practical work.

A better approach would be to make a complete *definitional conceptual schema*, i.e. the conceptual schema in which all concepts are completely defined and represented, and the users should have advanced interfaces to look at the schema in different ways. In collections of external or computerised concept descriptions of intensions of concepts should be stored and maintained, i.e. complete definitions are stored, whenever possible.

5 Information, Conceptual Constructs and Models

5.1 Stratified Description of the Human Cognition System and Its Information Content

In Section 2, we described the general characteristics of the stratified description of a system and identified 15 levels of human cognition. In it several notions of information and conceptual constructs can be identified. We need an opportunity to concentrate on one or a few types of constructs at a time, and we need an opportunity to work on one or a few levels at a time. Questions concerning this problem might be: what is the issue being described on this specific level, or what are the issues to be elicited, or what are the characteristics of model concepts for describing skill knowledge of riding a bicycle, or what are the characteristics of model concepts for describing a specific theory of criminology or economic values.

In Table 2 there are three columns: Levels of cognition, Levels of information and Levels of conceptual models. The first column lists 15 levels of cognition, which characterise levels of abstraction of constructs in human cognition. As explained in Section 2, there are principles between levels, which specify the rules in the level structure. The consequence of using this structuring principle is that the strata form a structure like a deck of cards in which the most abstract constructs are on the highest level, and the most concrete constructs are on the lowest level. The asymmetrical interdependence between the functioning of the system on different strata causes that the levels cannot in general be arranged in some other order. The other important aspect is that often a concept cannot be used on several levels.

In developing a structure like this it may happen that at first, two levels seem to be in one order, because the modeller is still learning the content of levels, but later, when she masters them, the levels seem to be in another order so that the former higher level is now the lower level. For example, when a person is practising shooting, she has to weigh or reason explicitly on several aspects of firing, and the level of skill knowledge seems to be higher than the level of skill itself, but later, when she already is the master, the level of skill knowledge seems to be lower than the level of skill of shooting itself.

The level of language (level 15) is here described as the highest level, and all conceptual levels (levels 3 – 14) are on lower levels. Two lowest levels are pre-conceptual, i.e. they contain sense data and elemental basics of information, which cannot as yet be recognised as mental images or even simple concepts.

The second column lists the corresponding levels of information, in which every level contains at least partly different notion of information, either because the concepts on that level are different from the concepts on the neighbouring levels, or because the rules on that level are different from the rules on the neighbouring levels, both on the lower level and the higher level. As above, on each level of information there are its own set of terms, concepts, and principles; there exists an asymmetrical

Table 2. Levels of cognition, levels of information and levels of conceptual models

Levels of cognition	Levels of information	Levels of conceptual models
Languages, and descriptions made by using the language ↑	Social information, pragmatic information, semantic information, syntactic information, discursive information	Conceptual description of a real or a hypothetical world, linguistic grammar, graphic grammar, language of the theory,
Values ↑	Value information	Value concepts, value structures, value ontologies
Norms ↑	Norm information	Norm concepts, norm structures, norm ontologies
Theories ↑	Laws, retrodictions, predictions, explanations, subsumptions, descriptions	Conceptual schemas, concept systems, information (rule) systems, verification systems
Cognitive models/conceptual schemas ↑	Structure of the cognitive model/conceptual schema	Conceptual schemas, concept systems
Art, skill, technique ↑	Competent use of skill	Art, skill, technique concepts, art or skill technique structures
Skill knowledge ↑	Complete skill knowledge	Skill knowledge concept, skill knowledge structures – skill descriptions
Inferencing mechanisms ↑	Discursive information, see Facts	Inferencing rule systems, problem specifications, rule system instances
Rules ↑	Discursive information, see Facts	Rule elements, rule schemas, rule instances
Symbolised instances ↑	Discursive information, see Facts	Concepts 1-3, symbols connected to concepts
Facts ↑	Discursive information, proceeding to a conclusion by reason or argument rather than intuition	Concepts 3
Perceptions of recognised instances ↑	Concursive information	Concepts 2, concepts and observations
Conceptual knowledge ↑	Concursive information	Concepts 1, mental models
Tacit knowledge ↑	Concursive information	Connectionistic models 2
Non-conceptualised observation data,	Implicit information	Connectionistic models 1

Symbol ↑ indicates the lowest level where the phenomenon mentioned on the same line can be identified. The same type of phenomena may also appear on higher levels, possibly in more complicated manifestations. Some levels may still be missing from this diagram. Concepts 1, 2, 3 are the most primitive sets of concepts.

interdependence between the functioning of a system on different strata, and the principles or laws used to characterise the system on any stratum cannot generally be derived from the principles used on other strata.

The third column lists the abstraction or strata levels of conceptual models. If we study the flow of information in conceptual models from one level to another, and start from the bottom of the multi-level structure and go on to higher levels, then on each level the conceptual model structure of that level, encompasses some new aspects, which originate partly from the abstract structuring principle of the strata structure, partly from concursive information received from adjoining information flows, and partly from discursive information delivered to adjoining information flows. Information exchange between adjoining information flows seems to depend on informational situations on conceptual levels.

On the lowest level there is only information that comes through the observation mechanisms. On this and on the second lowest level we may assume that information is represented as connectionistic constructs [43,44,45,46]. For simplicity, on the higher levels we assume that information is represented as structures constructed from concepts and relationships typical to each respective level [47]. In fact, in human brains all constructs are in the form of neural nets.

5.2 Accumulating Information as Possibly Actual Instances of Predefined or Potential Concepts

Next we study some aspects for collecting information from various sources in the UoD and accumulating it as possibly actual instances of various types of predefined or potential concepts. A *predefined concept* is a concept that has been constructed before the learning situation at hand. It is an ordinary concept, developed by thinking or combining existing knowledge primitives or concepts in previous learning or observation situations. Concepts may be e.g. individuals (particulars), universals, entities, stuffs, real kinds, events, relationships, continuants (i.e. endurants), occurrents (i.e. perdurants), or several other types of concepts [48,49,50,51].

For each concept (type) there is a characteristic set of properties, e.g. continuants are characterised as entities that are 'in time', i.e. they are 'wholly' present and all their proper parts are present at any time of their existence. Occurrents are entities that 'happen in time'. They extend in time by accumulating different 'temporal parts', so that, at any time t at which they exist, only some of their temporal parts at t are present. For example, the book you are holding now can be considered an endurant because (now) it is wholly present, while "your reading of this book" is a perdurant as your "reading" of the previous section is not present any more. [50]

An example is the notion of BUYING EVENT described briefly above. In it there are several temporal parts: TRADE NEGOTIATION, MAKING AN AGREEMENT, DELIVERY OF A COMMODITY, and PAYMENT. The continuant parts of the whole concept are the SELLER and the CUSTOMER, that are 'in time' all the time (or in office hours), and all others are temporal occurrents. TRADE NEGOTIATION starts when certain conditions are met, e.g. pre-conditions are met and an existence condition EC4 is also met. When the agreement has been made, or the customer has ceased negotiations, the trade negotiation is finished. In this situation, condition rule C3 fires and the next phase of the process will be started immediately (C3 is a temporal condition indicated by T in condition specification, and it activates a temporal operation *meet* (m)) [52]. Temporal condition C3 states that MAKING AN AGREEMENT is started immediately when TRADE NEGOTIATION has come to an end. Similarly, DELIVERY OF COMMODITIES is started immediately when MAKING AN AGREEMENT has been finished (existence condition EC7). PAYMENT is started immediately when MAKING AN AGREEMENT has been finished, and PAYMENT must end immediately when DELIVERY OF A COMMODITIES has been done. Finishing condition for ending the BUYING EVENT is FC8, i.e. when PAYMENT has been made, or TIME FOR PAYMENT has ended.

Natural information originating from the UoD comes to an observer either in small 'pieces', in larger amounts, or sometimes in very large masses, but never as concepts. A significant aspect is that sometimes the flow of information grows gradually so that first it is small and difficult to detect at all, later it grows larger and larger, until it finally comes in the form of some kind of an accident, a disaster, or a massive catastrophe. Usually a catastrophe does not come out of the blue, without any warnings, but some warning signs can be detected, if they or their origin can be properly recognised. The observer may receive concepts by reading or by recognising symbols, the content of which is some concept.

Information originating from the UoD is based on real objects and real events, as for example, on real cars involved in real accidents. This information has two important aspects: it has a *physical origin* and it is *true* in the sense that it is based on that physical origin. A problem appears when a reporter makes a piece of news about the accident that we saw earlier. He changes the type of information in writing a report about the incident. He no longer uses the original information – he is using some language. We cannot any longer be sure that the report is truthful. It depends on the text written by the reporter. The other problem is that false or erroneous messages may be sent with purpose, and the recipient has difficulties in recognising whether the message is correct or not.

Information in concepts seems to be at least partly of different origin and have different properties than information from real events. Only seldom is it based solely on real objects and real events. Concepts contain some aspects based on our own reasoning, or communication with other people, or sometimes only on speculation. On the other hand, concepts may contain aspects based on idealization, such as basic constructs in Euclidian geometry. Concepts may be abstracted, or they may contain abstracted parts, which may cause that by abstraction it is possible to construct a concept that does not have a physical counterpart in the real world (as e.g. value concepts), or it is possible to construct a concept by abstraction from an abstract concept [53].

These types of information can be separated: the first one is *instance information* and the other is *concept information*. The concept information describes the structure and type information that this concept contains. Instance information tells us whether the instance has occurred or not, and also more or less detailed perceptions of the event. Both types of information must be consistent with each other. Learning may be based on both types of information, or only one of them.

5.3 Example

Let us suppose that a car accident is going to happen. We have a pre-defined concept of a car accident, or a set of concepts of different types of car accidents in our minds and also stored in the information system. We may also have many other types of pre-defined concepts. In the distributed information system there is a mechanism that can collect instance information of various events in a wide area, e.g. a region of a town. The system can also store large amounts of instance information of events occurred earlier and accepted as valid instances of car accidents, or other events. Both systems are connected so that the collection of concept information and corresponding occurrences of instance information are linked together. If the distributed information system recognises sounds or/and pictures of an accident, it analyses the situation, and on the basis of that, possibly sends an ambulance and/or a police patrol on the site.

This example shows a quite simple, but advanced system for keeping track of certain types of situations in the area. More difficult cases may be situations in which we do not yet know any examples of real instances but should be prepared for them. We can construct several high level pre-defined concepts and several different types of pre-defined instance information. If some new type of event appears, we can first check these pre-defined instance information sequences and after that we can start constructing combinations of different pre-defined instance information sequences. The other case is that we get some small advance hints of coming phenomenon, but we do not yet know what it is.

For these situations we could develop many different conceptual models of cases that we should be able to manage.

First, we may have a concept of a person who has some general properties (Fig. 7A). We may have also specialised types of people, e.g. an employee, a policeman, a taxi-driver, and a senior citizen, that may appear in some events (Fig. 7B). A taxi is a concept that contains a concept of a taxi-driver (a person) and a concept of a cab (a car), which both serve a concept of a customer (Fig. 7C). A taxi-owner is a person who owns and maintains one or more taxicabs, driven by taxi-drivers. Fig. 7D shows a general conceptual model for describing an event of one or more (at most 4) customers (persons) riding by a taxi from one place to another in the town. Each customer may have his own place or address in where he climbs into the cab and also in where he is departing the cab. He can enter or leave out the cab either in some street-address or in some crossing of streets.

Fig. 7E shows a general conceptual model describing an event of an accident between a person and a cab. It is described by using entities and relationships, and also some continuants and several sub-events and time sequences between sub-events that are described as occurrents. Continuants are e.g. persons, sites, streets and street-crossings and a car. In addition, also the reason of the accident, the time of the accident and the guilty of the accident are continuants, although the instances of these concepts do not exist before the accident. These instances are created either at the time of the accident (TIME OF ACCIDENT) or usually shortly after it (GUILTY and REASON), but in some cases even shortly before the accident. The instance of the concept of the time of the accident consists of two parts: the first part contains an occurrent, which contains the beginning time of the accident and its ending time. The second part contains a continuant, which also contains the beginning time of the accident and its ending time, but because this concept is a continuant, it retains the values of both components.

In Fig. 7E there are some constraints, CR1, CR2, CR3 and CR4, which define the relationships between perceptions and time. In this case only the relationships between the time of the accident and visual observations are given. Constraint CR1

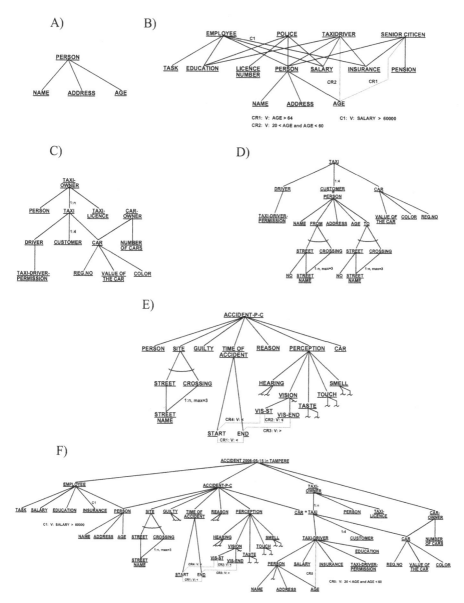

Fig. 7. Conceptual description of a sudden pedestrian-taxi accident

states that value of the start time of the accident must be less than the end time. Constraint CR2 states that value of the visual start time of the accident must be less than the visual end time. Constraint CR3 states that the value of the end time of the accident must be greater than the visual end time. Constraint CR4 states that the value of the start time of the accident must be less than the visual end time.

In Fig. 7F, the conceptual model of an accident between an employee and a taxicab in Tampere, on May 15, 2006 is described. The model is quite detailed, but it does not so far contain instance information. It contains some slots for instance information. The detailed instance information is available only when the accident is happening; and afterwards there are only some mental images stored in the memory of casual witnesses, and possibly instance information stored into the distributed information system, described briefly above. Some instance information may be collected from the remaining traces of the accident.

The example can be developed further in several directions, e.g. by studying whether the accident was really coincidental or was it purposeful. How can the police find it out?

Only some concepts, which fit to this collected information may be constructed automatically from instances. We have to remember, that the intension of a concept determines extensions, because the intension is based on properties of the concept, but not conversely. Therefore instance information cannot define concepts, except in the case that a concept is defined to contain exactly the same properties, which can be recognised on the instance information. But how can we determine which properties to select. We will not take into account here questions concerning possibilities of applying methods of machine learning and data mining.

Some concepts can be constructed after the incident by using automatically collected instance information, information received from the studies made after the accident, and from official or scientific investigations. The construction of new concepts and receiving new instance information makes learning possible. That learning may first contain the results of the official investigation, which describes the reason of the accident, answer to the question who was guilty for that accident, what was happening, and what were the consequences. Later, additional learning may happen if additional scientific investigations are made.

A problem to be still analysed more in detail is that usually predefined or potential temporal continuants or occurrents in the conceptual model cannot cover in advance all alternatives that may appear in instance information of the actual perdurant events. These instances may be cases of sudden events or processes. Nor is it likely that we could handle all different temporal sequences of events in advance. We must therefore be prepared to develop conceptual and/or statistical prediction and estimation methods, which facilitate 'reasonable precaution' in situations that cannot be evaluated in advance or anticipated.

Conceptual prediction is a method in which, by constructing new concepts and calculation methods, the consequences of using these new concepts in or against events to be studied, will be evaluated or developed in a real or virtual world. By using these methods we try to find concepts and technical implementations of them, which have an intended effect to or against events to be studied, in other words, we may need many simulation mechanisms.

6 Conceptual Modelling, Understanding and Learning

In this section we analyse various perspectives, which must be taken into account in studying conceptual modelling, human understanding and learning in the context of active conceptual modelling. These three notions and their relationships form a complex structure of human knowledge, which is difficult to split into separate parts. Conceptual modelling is based on many aspects of information, conceptual content and structure of concepts and conceptual models, and on skills of the modeller and users, as discussed in Chapter 3.

Learning as the acquisition of new knowledge and skills is generally regarded as a constructive activity, which first is based on gathering new information from the UoD, either by exchanging messages or collecting information from various sources. A learner usually acquires new information and skills in small portions. Often these portions are not well defined and therefore learning may be time consuming and errors frequent. However, sometimes learning happens in sudden and well-organised steps. Acquisition of information may result, that the learner gets first *immediate information*, i.e. new knowledge included into the messages or collected information, and then *derived information*, i.e. knowledge constructed by reasoning from new information and receivers own knowledge.

In analysing learning there are still difficult problems open. One problem is that learning is very many-sided phenomenon. There are a lot of different theories about the nature of learning and about methods and basic constructs of learning, see e.g. [54]. Usually several theories can be applied but often with different results. We will not go into details of these problems here.

D. Landy and R. L. Goldstone published a paper about the situation: "How we learn about things we don't already understand" [55]. This situation is common e.g. in astronomy and in solving crimes. It may appear in other situations, too. In astronomy, a researcher may encounter an observation that is completely new and unknown to all astronomers. A detective may find himself in a situation, in which he must first study whether a crime has happened or not. Then he must study what has happened and who is the victim and who is the criminal?

In a usual situation a learner must first study what is the problem he has to solve, which kind of information he needs, and how he can get it. He must study the material well, and organise new knowledge in a systematic way and integrate it into his own knowledge structures and to his own human conceptual schema. Integration of new knowledge to the conceptual schema is often an essential phase, which can be supported by external concept mapping tools [56], or with an advanced conceptual schema facility. The best result can be achieved if the modeller (or the user) can use the conceptual schema facility first to support for organising his own human conceptual schema as well as possible, and then use that facility for organising the external conceptual schema in such a way that both schemata are as much as possible isomorphic.

Learning and understanding are closely related activities in which understanding organises the conceptual schema in a new way. M. Bunge says that understanding a thing means knowing (in a deep sense) the position and role of the thing in the environmental epistemic framework into which it is embedded [57], and also knowing the internal structure and behaviour of it (or the model of it). The epistemic framework, i.e. concepts and rules, cannot only form an agglomeration of knowledge. It must be organised into a systematic theory based on concepts and rules of the corresponding discipline. These concepts and rules must form a conceptual system, which is related to the thing to be understood, e.g. in learning [57].

The resulting conceptual system should be organised as well as possible, taking into account the goals of the system, the goals of the user, and all the concepts applied, and their usage in thinking. According to Bunge, understanding is not an all-or-none operation: it comes in several kinds and degrees. Understanding calls for answering to what is called the six w's of science and technology: what (or how), where, when, whence, whither, and why. In other words, understanding is brought about description, subsumption, explanation, prediction, and retrodiction [58].

Bunge says that neither of these epistemic operations is more important that the others, although many researchers specialize in only some of them. Subsumption, explanation, prediction and postdiction are typical theoretical operations, i.e. activities performed by means of theories – the better, the richer and the better organized. [57].

If we receive some information based on a piece of the real object not in concert with the conceptual schema, that incoming information is disturbing the conceptual schema and our understanding of that conceptual schema. Then there are three alternative ways of acting: either we reject the incoming information, we try to adjust the conceptual schema to include the incoming information, or we try to adjust both the conceptual schema and the incoming information. Adjusting can be a complicated process, because it may depend on ontology used in constructing the conceptual schema, it may depend on ontology used in constructing the incoming information, or it may depend also on the epistemological theory of the conceptual schema, or several other aspects (see Chapters 3 and 4.2).

7 Discussion

This work is only the beginning of a much more extensive effort, but we can already recognise several aspects in which there are promising possibilities to develop working systems, e.g. accident recognition systems or other systems in which information stored in the conceptual schema can be compared to instance information and on the basis of that comparison some action is made. On the other hand, the question of the ability of the system to make extensive inferences from the difference between the conceptual schema and instance information is as yet unresolved.

References

1. van Griethuysen, J.J. (ed.): Concepts and Terminology for the Conceptual Schema and the Information Base. ISO/TC97/SC5-N695, 1982, 1990, ISO Secreteriat, Geneve
2. Mesarovic, M.D., Macko, D., Takahara, Y.: Theory of Hierarchical, Multilevel Systems. Academic Press, London (1970)
3. Nauta Jr., D.: The Meaning of Information. Mouton & Co., The Hague (1972)
4. Senko, M.E., Altman, E.B., Astrahan, M.M., Fehde, P.L.: Data Structures and Accessing in Data Base Systems. I Evolution of Information Systems. II Information Organization. III Data Representation and the Data Independent Accessing Model. IBM System Journal 12(1), 30–93 (1973)
5. Kangassalo, H.: A Stratified Description as a Framework of File Description. (Kerrostettu kuvaus tiedostojen kuvauksen kehysmallina). DIFO-vuosikirja 2/1973. DIFO-tutkimus ry, Tampere, pp. 47–62 (May 1973) (in Finish)

6. Kangassalo, H.: Conceptual Structure of the File Design Problem: a General View. DIFOreport no 8. In: Bubenko Jr., J.A., Sølvberg, A. (eds.) Data Base Schema Design and Evaluation, p. 18, Røros, Norway (1975-02-27...28)

7. Kangassalo, H.: Structuring Principles of Conceptual Schemas and Conceptual Models. In: Bubenko Jr., J.A. (eds.) Information Modelling, pp. 223–311, Studentlitteratur, Lund (1983)

8. Nauta Jr., D.: The Meaning of Information, p. 170. Mouton & Co, The Hague (1972)

9. FRISCO The IFIP WG 8.1 Task Group FRISCO A Framework of Information System Concepts, p. 54 (December 1999)

10. Nauta Jr., D.: The Meaning of Information, p. 169. Mouton & Co, The Hague (1972)

11. Lacey, A.R.: A Dictionary of Philosophy. Routledge & Kegan Paul, London (first published 1976)

12. Flavell, J.H., Miller, P.H., Miller, S.A.: Cognitive development, 3rd edn., p. 2. Prentice-Hall, Englewood Cliffs (first ed. 1977) (1993)

13. Bermúdez, J.: Nonconceptual Mental Content. Stanford Encyclopedia of Philosophy. Online edition 2005-08-31

14. Polanyi, M.: The Tacit Dimension. Routledge & Kegan Paul, London (1966)

15. Polanyi, M.: Personal Knowledge. Towards a Post-Critical Philosophy. Routledge & Kegan Paul, Ltd., London (First published 1958) (2002)

16. Wiig, K.M.: Knowledge Management Foundations: Thinking about Thinking, p. 11. Schema Press, Arlington, Texas (1993)

17. Kangassalo, H.: COMIC - A System and Methodology for Conceptual Modelling and Information Construction. Data & Knowledge Engineering 9, 287–319 (1992/1993)

18. Guarino, N., Giaretta, P.: Ontologies and Knowledge Bases: Towards a Terminological Clarification. In: Mars, N.J.I. (ed.) Towards Very Large Knowledge Bases, IOS Press, Amsterdam (1995)

19. Guarino, N.: Formal Ontology in Information Systems, pp. 3–15. IOS Press, Amsterdam (1998)

20. Bilgrami, A.: Rules. In: Honderich, T. (ed.) The Oxford Companion to Philosophy, p. 781. Oxford University Press, Oxford (1995)

21. Pettit, P.: Problem of rule-following. In: Dancy, J., Sosa, E., (eds.) A Companion to Epistemology. Blackwell Companion to Philosophy. Blackwell Publishers, Inc., Oxford, UK (1996) (Rule-following using words requires that a language and appropriate words are available. See also Point 14. Norms)

22. White, J.: Inference. In: Honderich, T. (ed.) The Oxford Companion to Philosophy, Oxford University Press, Oxford (1995)

23. Ruse, M.: Theory. In: Honderich, T. (ed.) The Oxford Companion to Philosophy, Oxford University Press, Oxford (1995)

24. Stegmüller, W.: The Structure and Dynamics of Theories. Springer, Berlin (1976)

25. Stegmüller, W.: The Structuralist View of Theories. In: A Possible Analogue of the Bourbaki Programme in Physical Science. Springer, Berlin (1979)

26. Weitz, M.: Theories of Concepts. In: A History of the Major Philosophical Tradition. Roudledge, London (1988)

27. Bunge, M.: Treatise on Basic Philosophy. In: Semantics I: Sense and Reference, vol. 1, D. Reidel Publishing Company, Dordrecht, Holland (1974)

28. Laurence, S., Margolis, E.: Concepts and Cognitive Science. In: Margolis, E., Laurence, S. (eds.) Concepts - Core Readings, MIT Press, London (1999)

29. Murphy, G.L.: The Big Book of Concepts. A Bradford Book. The MIT Press, Cambridge (2002)

30. Bunge, M.: Characterises epistemology as follows: Epistemology, or the theory of knowledge, is the field of research concerned with human knowledge in general-ordinary and scientific, intuitive and formal, pure and action-oriented. And methodology-not to be mistaken for methodics, or a set methods or techniques-is the discipline that studies the principles of successful inquiry, whether in ordinary life, science, technology, or the humanities, Bunge M, Treatise of Basic Philosophy, vol. 5, p. xiv

31. Klausmaier, H.J.: Conceptualizing. In: Jones, B.F., Idol, L. (eds.) Dimensions of Thinking and Cognitive Instruction, pp. 93–138. Lawrence Erlbaum, Hillsdale, N.J. (1990)

32. Palomäki, J.: From Concepts to Concept Theory. Discoveries, Connections, and Results. PhD Diss. Acta Universitatis Tamperensis. Ser. A, vol. 416, University of Tampere, Tampere (1994)

33. Moser, P.K.: Epistemology. In: Audi, R. (General Editor): The Cambridge Dictionary Philosophy, Cambridge University Press, Cambridge (1996)

34. Medin, D.L., Goldstone, R.L.: Concepts. In: Eysenck, M.W. (ed.) The Blackwell Dictionary of Cognitive Psychology, pp. 77–83. Basil Blackwell, Oxford, UK (1990)

35. Margolis, E., Laurence, S. (eds.): Concepts - Core Readings. MIT Press, London (1999)

36. Kelly, G.H.: The Psychology of Personal Constructs. W.W. Norton & Company, New York, Two volumes (1955)

37. Thagard, P.: Conceptual Revolutions, pp. 34–39. Princeton University Press, Princeton (1992/1993)

38. We will assume that ontology is a conceptual system, in which all concepts are basic or primitive concepts, from which all other, more advanced concepts can be constructed. A concept is a basic or primitive concept, if it does not contain any concepts contained into it

39. Kangassalo, H.: Dynamics of a Conceptual Schema and its Description. Department of Computer Sciences, University of Tampere (2007-01-09) (under preparation)

40. Kauppi, R.: Einführung in die Theorie der Begriffssysteme. In: Acta Universitatis Tamperensis, Ser. A, vol.15, Universität Tampere, Tampere (1967)

41. Kangassalo, H.: Frameworks of Information Modelling: Construction of Concepts and Knowledge by Using the Intensional Approach. In: Brinkkemper, J., Lindengrona, E., Sølvberg, A. (eds.) Information Systems Engineering. State of the Arts and Research Themes, Springer, Heidelberg (2000)

42. Rumelhart, D.E., Hinton, G.E., McClelland, J.L.: A General Framework for Parallel Distributed Processing and The PDP Research Group. In: Parallel Distributed Prosessing. Explorations in the Microstructure of Cognition, vol. 1, ch. 2, A Bradford Book, The MIT Press, Cambridge, Massachusetts (1988) (Eigth printing)

43. Hinton, G.E., McClelland, J.L., Rumelhart, D.E.: Distributed Representations. In: Rumelhart, D.E., et al. (eds.) Parallel Distributed Prosessing. Explorations in the Microstructure of Cognition. vol. 1: Foundations, ch. 3 (see above)

44. Rumelhart, D.E., McClelland, J.L.: PDP Models and General Issues in Cognitive Science. In: Rumelhart, D.E., et al. (eds.) Parallel Distributed Prosessing. Explorations in the Microstructure of Cognition. vol. 1: Foundations, ch. 4 (see above)

45. Smolensky, P.: Information Processing in Dynamical Systems: Foundations of Harmony Theory. In: Rumelhart, D.E., et al. (eds.) Parallel Distributed Prosessing. Explorations in the Microstructure of Cognition. vol. 1: Foundations, ch. 6 (see above)

46. It would be natural to assume, that also on higher levels the constructs are connectionistic constructs, but representations of them would be complex and large. We simplify representations and say that they are constructed from concepts and relationships typical to each respective level. We assume that they can be reduced to connectionistic constructs

47. Medin, D.L., Lynch, E.B., Solomon, K.O.: Are There Kinds of Concepts? Annu. Rev. Psychol. 51, 121–147 (2000)
48. Millikan, R.G.: A Common Structure for Concepts of Individuals, Stuffs, and Real Kinds: More Mama, More Milk and More Mouse. Behavioral and Brain Sciences 22(1), 55–65 (1998)
49. Masolo, C., Borgo, S., Gangemi, A., Guarino, N., Oltramari, A.: WonderWeb Deliverable D18. Ontology Library (final). Laboratory For Applied Ontology- ISTC-CNR, Trento, Italy (2003)
50. Chisholm, R.M.: A Realistic Theory of Categogies-An Essay on Ontology. Cambridge University Press, Cambridge (1996)
51. Allen, J.F.: Maintaining Knowledge about Temporal Intervals. Comm. of the ACM 26(11) (November 1983)
52. Kangassalo, H.: On the Concept of Concept for Conceptual Modelling and Concept Detection. In: Ohsuga, S., Kangassalo, H., Jaakkola, H., et al. (eds.) Information Modelling and Knowledge Bases III, pp. 17–58. IOS, Amsterdam (1992)
53. Emerging Technologies: Learning Theories (2007-05-17) http:// www. emtech. net/ learning_theories.htm
54. Landy, D., Goldstone, R.L.: How we learn about things we don't already understand. Journal of Experimental & Theoretical Artificial Intelligence 17(4), 343–369 (2005)
55. Novak, J.D., Cañas, A.J.: Building on New Contructivist Ideas and Cmap Tools to Crate a New Model for Education. In: Cañas, A.J., Novak, J.D., González, F.M. (eds) Concept maps: Theory, methodology technology. Proc. of the 1st Int. Conf. on Concept Mapping. Universidad Pública de Navarra, Pamplona, Spain (2005)
56. Bunge, M.: Treatise on Basic Philosophy. In: Epistemology & Methodology II: Understanding the World, vol. 6, D. Reidel, Dordrecht, Holland (1983)

Spatio-temporal and Multi-representation Modeling: A Contribution to Active Conceptual Modeling

Stefano Spaccapietra[1], Christine Parent[2], and Esteban Zimányi[3]

[1] Database Laboratory, Ecole Polytechnique Fédérale de Lausanne,
CH-1015 Lausanne, Switzerland
Stefano.Spaccapietra@epfl.ch
[2] HEC ISI, University of Lausanne,
CH-1015 Lausanne, Switzerland
Christine.Parent@unil.ch
[3] Department of Computer & Decision Engineering (CoDE), Université Libre de Bruxelles,
50 av. F.D. Roosevelt, 1050 Bruxelles, Belgium
ezimanyi@ulb.ac.be

Abstract. Worldwide globalization increases the complexity of problem solving and decision-making, whatever the endeavor is. This calls for a more accurate and complete understanding of underlying data, processes and events. Data representations have to be as accurate as possible, spanning from the current status of affairs to its past and future statuses, so that it becomes feasible, in particular, to elaborate strategies for the future based on an analysis of past events. Active conceptual modeling is a new framework intended to describe all aspects of a domain. It expands the traditional modeling scope to include, among others, the ability to memorize and use knowledge about the spatial and temporal context of the phenomena of interest, as well as the ability to analyze the same elements under different perspectives. In this paper we show how these advanced modeling features are provided by the MADS conceptual model.

Keywords: Active conceptual models, spatio-temporal information, multiple representations, multiple perspectives, MADS model.

1 Introduction

Globalization has significantly increased the complexity of the problems we face in many different areas, e.g., economic, social, and environmental. Events and phenomena have intricate and subtle interdependencies whose perception changes when seen from different perspectives. This calls for more knowledgeable data management systems, capable of managing a more comprehensive description of the world. Knowledge of cause-effect relationships between events, for example, contributes to the evaluation of the future impact of solutions to problems in a decision-making environment. Analyses of past events and phenomena supports

P.P. Chen and L.Y. Wong (Eds.): ACM-L 2006, LNCS 4512, pp. 194–205, 2007.
© Springer-Verlag Berlin Heidelberg 2007

decision makers with the possibility to learn from past experiences and to take them into account in choosing future actions. This is a very traditional learning pattern, but poorly supported by current data management systems.

More knowledgeable systems are the target of active modeling, a paradigm defined as a continual process describing the important and relevant aspects of the real world, including the activities and changes, under different perspectives [1]. It aims at providing control and traceability for the evolving and changing world state, helping to understand the relationships among changes. An active conceptual model provides at any given time a multilevel and multi-perspective high-level abstraction of reality. Consequently, one of its basic features is integrating time and space modeling[1].

Unfortunately, the current established conceptual modeling practices do not achieve these objectives. They are still limited to mainly deal with organizing basic data structures, e.g. describing entities, relationships, properties and possibly processes. Other directions emerge that slowly influence progress in DBMS development. Supporting the description of spatial features has recently become part of the functionality of most recent DBMS, although to a still rather primitive extent. Temporal information is most frequently interleaved with spatial information, but is very poorly supported, which entangles the development of a multitude of applications that have to deal with spatio-temporal data, e.g. moving objects.

In this paper we show how spatial and temporal characteristics of phenomena can be captured into a conceptual model. We also discuss how to provide multiple representations of phenomena, thus allowing to analyze them under different perspectives, e.g., with respect to different stakeholders. Concrete modeling concepts are proposed using the constructs provided by the MADS conceptual model [2]. For readers already familiar with MADS, the paper offers a short presentation of the major features of the model.

2   Conceptual Modeling: The MADS Approach

Conventional conceptual models and DBMS have been tailored to manage a static view of the world of interest. They capture the state of affairs at a given moment in time (in the terminology of the temporal database community, they capture a snapshot). Active modeling stems from a dynamic view of the world as something that is continuously changing and where knowledge about the changes is as important as knowledge about the current status. This view entails their focus on capturing discrete as well as continual changes (i.e. time-varying information) and cause-effect relationships that help in understanding these changes. Since work on temporal databases, temporal aspects have evolved into spatio-temporal aspects, which capture phenomena that vary in both space and time. Moving objects (e.g. people, parcels, cars, clouds) are typical examples of spatio-temporal objects. Thanks to the availability of mobile devices, e.g. GPS, and ubiquitous computing, interest in mobile objects applications has exploded in recent years. Active modeling appears as the natural evolution we need today to face such innovative applications.

---

[1] cf. the Call for Papers of the First International Workshop on Active Conceptual Modeling of Learning (ACM-L 2006), Tucson, Arizona, November 8, 2006.

Finally, conventional models poorly support different and multiple perspectives of the same real-world phenomena, another characteristic of modern applications that are ruled by decentralized paradigms. Decentralization entails diversity and complementarities. There is not anymore a single enterprise-wide apprehension of data. Instead, the flexibility offered by applications handling multiple perspectives is a decisive advantage towards better strategies and decision-making processes.

The MADS model has been defined as an answer to the above requirements. MADS [2,3] is indeed a conceptual spatio-temporal data model with multi-representation support. It handles its four modeling dimensions - structural, spatial, temporal, and multi-representation – in a way that purposely makes the modeling dimensions *orthogonal* to each other (i.e., modeling in one dimension is not constrained by modeling choices in another dimension). Consequently, MADS can also be efficiently used in simpler environments, such as classical non-spatial, non-temporal, and mono-representation databases.

In terms of the framework for active conceptual modeling proposed in [4], MADS model covers several of the areas stated as necessary. It provides a comprehensive approach for *multi-level and multi-perspective modeling*, where multiplicity of level and perspectives can be handled at both the schema and instance levels. This guarantees maximum flexibility and precise match with specific application requirements. MADS allows both to customize the view of the system provided to different users, as well as cope with the inevitable inconsistencies and incompatibilities that arise when providing different perspectives of the same information. More precisely, MADS allows inconsistencies to exist if needed, btu guarantees that each perception is consistent if takes in isolation. *Space and time* features are smoothly integrated into the conceptual model allowing designers to precisely define how knowledge about where and when the events of interest occurred has to be kept. MADS aims to be an *executable conceptual model* in two different ways. First, it provides an associated query and manipulation language at the conceptual level, thus covering the full lifecycle of the information system in a uniform approach. Second, the conceptual specifications, whether for defining the schema or manipulating the database, are translated through CASE tools into the particular language provided by current implementation platforms (e.g., SQL) [3]. These tools provide users with visual interaction functionality, relieving them from textual languages and logical-level concerns. Finally, MADS definition language as well as its query and manipulation languages are formally defined.

MADS is not the only conceptual spatio-temporal model that has been described in the literature. The work reported in [5] also offers good support for spatio-temporal modeling, but does not address multiplicity of perspectives. Many more spatio-temporal data models are reported in [6]. Perceptory is the only spatio-temporal model that also proposes a mechanism for supporting multiple perspectives [7]. This mechanism is close to traditional database view techniques and does not have the full functionality MADS offers.

In the following sections we briefly describe some of the main characteristics of the four modeling dimensions of MADS, namely, structural, spatio-temporal, and multi-representation. We use as a running example a small excerpt of a real-word application coping with natural risk management in mountainous regions, e.g., avalanches and landslides (cf. Fig. 1). A full description of MADS is available in [2].

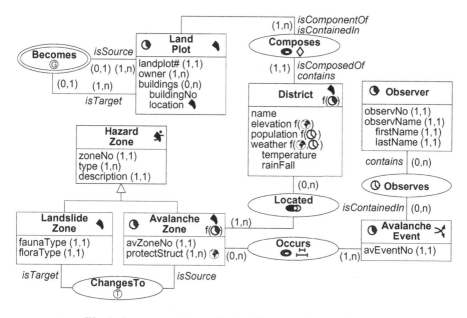

Fig. 1. An excerpt of an application for managing natural risks

3 Structural Modeling

MADS provides rich mechanisms for describing the data structure component of applications. In the description that follows we concentrate on novel aspects of MADS with respect to traditional conceptual models.

MADS structural dimension includes well-known features such as objects, relationships, attributes, and methods. Objects and relationships have an identity and may bear attributes. Attributes are mono-valued or multi-valued, simple or complex (i.e., composed of other attributes), optional or mandatory, and may be derived.

MADS identifies two basic kinds of relationship types, association and multi-association. An *association* is the typical kind of relationship type where each of its roles links one and only one instance of the linked object type. However, in some situations the basic association relationship does not allow to accurately represent real-word links existing between objects. For example, due to territorial reorganization, one or several land plots may be split/merged into one or several land plots. Fig. 1 shows a multi-association (drawn with a double oval) Becomes. This multi-association allows setting up a direct link between the set of land plots that is the input of the reorganization process and the set of land plots that is the output. Consequently, each role in a multi-association relationship type bears two pairs of (minimum, maximum) cardinalities. A first pair is the conventional one that, as in association relationship types, defines for each object instance, how many relationship instances it can be linked to via the role. The second pair defines for each relationship instance, how many object instances it can link with this role. Its value for minimum is at least 1. Obviously, an association is nothing but a special case of multi-association (with all maxima for the second cardinality pairs equal to 1). Pragmatic reasons (simplicity, frequency of use, user familiarity with the concept) make it nevertheless worth having associations as a separate construct.

Semantic data models usually provide the possibility to link objects through different types of relationships, each one with a specific semantics. MADS, instead, separates the definition of relationships into two facets. First, the appropriate link is built using either the association or the multi-association construct (this is the structural component of the link). Second, whenever needed, the link is given one or more specific semantics that convey the semantic implications of the link. MADS supports aggregation, generation, transition, topological, synchronization, and inter-representation semantics for the relationships. *Aggregation* is the most common one: It defines mereological (also termed component or part-of) semantics. An example is the relationship Composes in Fig. 1 (identified by the ◊ icon). *Generation* relationships record that one or several target objects have been generated by other source objects. An example in Fig. 1 is the Becomes relationship. Finally, *transition* semantics expresses that an object in a source object type has evolved to a new state that causes it to be instantiated in another target object type. For example, whenever efficient protection structures are built in an avalanche zone, the zone is no longer regarded as an avalanche zone but may become a zone subject to landslide: The instance moves from the AvalancheZone object type to the LandslideZone object type. If the application needs recording this change of classification, the ChangesTo transition relationship type (identified by the ⓣ icon) can be added to the schema, as shown in Fig. 1. Other possible semantics for relationships are defined in the following subsections, to deal with spatial, temporal, and multi-representation features.

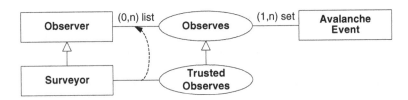

Fig. 2. A relationship subtype refining a role to link a subtype of the original object type

The generalization/specialization (or is-a) relationship allows relating a generic type, the supertype (e.g., HazardZone in Fig. 1) and a specific one, the subtype (AvalancheZone in Fig. 1) stating that they convey different representations of the same real-world phenomena. MADS supports is-a links between relationship types, with the same characteristics as between object types. In Fig. 2, the designer wants to create a separate relationship type for observations of avalanches made by surveyors, since any person may be introduced in the database as reporting an avalanche. A sub-relationship type may have additional properties and additional roles with respect to the super-relationship type. Further, as shown in the figure, existing roles may be refined as associated to a subtype of the otherwise inherited object type. The design ensures that only observations made by surveyors are registered as instances of TrustedObserves.

4 Spatio-temporal Modeling

In MADS space and time description is orthogonal to data structure description, which means that the description of a phenomenon may be enhanced by spatial and temporal features whatever data structure (i.e., object, relationship, attribute) has been chosen to represent it. MADS allows describing spatial and temporal features with either a discrete or a continuous view. These are described next.

The *discrete view* (or *object view*) of space and time defines the spatial and temporal extents of the phenomena of interest. The *spatial extent* is the set of 2-dimensional or 3-dimensional points (defined by their geographical coordinates <x,y> or <x,y,z>) that the phenomenon occupies in space. The *temporal extent* is the set of instants that the phenomenon occupies in time. Temporality in MADS corresponds to *valid time*, which conveys information on when a given fact, stored in the database, is considered valid from the application point of view.

Specific data types support the definition, manipulation, and querying of spatial and temporal values. MADS supports two hierarchies of dedicated data types, one for spatial data types, and one for temporal data types. Generic spatial (resp. temporal) data types allow describing object types whose instances may have different types of spatial extents. For example, a River object type may contain large rivers with an extent of type Surface and small rivers with an extent of type Line. Examples of spatial data types are: Geo (⊕), the most generic spatial data type, Surface (◥), and SurfaceBag (◢). The latter is useful for describing objects with a non-connected surface, like an archipelago. Examples of temporal data types are: Instant (ⵔ), Interval (◉), and IntervalBag (✪). The latter is useful for describing the periods of activity of non-continuous phenomena.

A *spatial* (*temporal*) *object type* is an object type that holds spatial (temporal) information pertaining to the object itself. For example, in Fig. 1 Landplot is both a spatial and temporal object type as shown by the Surface (◥) and the Interval (◉) icons on the right and left sides of the object type name. Following common practice, we call *spatio-temporal* an object type that either has both a spatial and a temporal extent, separately, or has a time-varying spatial extent, i.e., its spatial extent changes over time and the history of extent values is recorded (e.g., District and AvalancheZone in Fig. 1). Time and space-varying attributes are described hereinafter. Similarly, *spatial*, *temporal*, and *spatio-temporal relationship types* hold spatial and/or temporal information pertaining to the relationship as a whole, exactly as for an object type. For example, in Fig. 1, the Observes relationship type is temporal, of kind Instant, to record when observations are made.

The spatial and temporal extents of an object (or relationship) type are kept in dedicated system-defined attributes: geometry for the spatial extent and lifecycle for temporal extent. geometry is a spatial attribute (see below) with any spatial data type as domain. When representing a moving or deforming object (e.g. AvalancheZone), geometry is a time-varying spatial attribute. On the other hand, the lifecycle allows database users to record when, in the real world, the object (or link) was (or is planned to be) created and deleted. It may also support recording that an object is temporarily suspended, like an employee who is on temporary leave. Therefore the lifecycle of an instance says at each instant what is the status of the corresponding real world object (or link): scheduled, active, suspended, or disabled.

A *spatial (temporal) attribute* is a simple attribute whose domain of values belongs to one of the spatial (temporal) data types. Each object and relationship type, whether spatial, temporal, or plain, may have spatial, temporal, and spatio-temporal attributes. For example, in Fig. 1 the LandPlot object type includes, in addition to its spatial extent, a complex and multivalued attribute buildings whose second component attribute, location, is a spatial attribute describing, for each building, its spatial extent.

Practically, the implementation of a spatial attribute, as well as the one of a geometry attribute, varies according to the domain of the attribute. For instance, a geometry of kind Point (in 2D space) is usually implemented by a couple of coordinates <X,Y> for each value, a geometry of kind Surface by a list of couples <X,Y> per value, and a geometry of kind SurfaceBag by a list of lists of couples <X,Y> per value.

Constraining relationships are binary relationships linking spatial (or temporal) object types stating that the geometries (or lifecycles) of the linked objects must comply with a spatial (or temporal) constraint. For example, in Fig. 1 Composes is an aggregation and also a constraining relationship of kind topological inclusion, as shown by the ● icon. The constraint states that a district and a land plot may be linked only if the spatial extent of the district effectively contains the spatial extent of the land plot. Relationship types may simultaneously bear multiple semantics. For example, Occurs is both a topological and synchronization constraining relationship type. It ensures that both spatial and temporal extents of AvalancheZone are defined such that every avalanche event always occurs inside (topological constraint ●) a currently existing (synchronization within constraint ⊐⊏) avalanche zone.

Beyond the discrete view, there is a need to support another perception of space and time, the *continuous view* (or *field view*). In the continuous view a phenomenon is perceived as a function associating to each point (or instant) of a spatial (or temporal) extent a value. In MADS the continuous view is supported by space (and/or time) *varying attributes*, which are attributes whose value is a function that records the history – and possibly the future – of the value. The domain of the function is a spatial (and/or temporal) extent. Its range can be a set of simple values (e.g., Real for temperature, Point for a moving car), a set of composite values if the attribute is complex as, in Fig. 1, weather, or a powerset of values if the attribute is multivalued.

District in Fig. 1 shows examples of varying attributes and their visual notation in MADS (e.g., f(⬤)). elevation is a space-varying attribute defined over the geometry of the district. It provides for each geographic point of the district its elevation. population is a time-varying attribute defined over a constant time interval, e.g. [1900-2006]. weather is a space and time-varying complex attribute which records for each point of the spatial extent of the district and for each instant of a constant time interval a composite value describing the weather at this location and this instant. Such attributes that are space and time-varying are also called *spatio-temporal attributes*. The geometry attribute can also be time-varying, like any spatial attribute. For instance in Fig. 1, both AvalancheZone and District have a time-varying geometry. The deformation of their spatial extent can therefore be recorded.

Practically, the implementation of a continuous time-varying attribute is usually made up of 1) a list of <instant, value> pairs that records the sample values, and 2) a method that performs linear interpolation between two sample values. For instance, a time-varying point would be implemented by a list of triples <instant, X,Y>. On the other hand, the deforming surface of District, which is time-varying in a step-wise manner, would be implemented by a list of couples <time interval, LIST(<X,Y>)>.

A constraining topological relationship may link moving or deforming objects, i.e., spatial objects whose geometries are time-varying. An example in Fig. 1 is the relationship Located linking District and AvalancheZone, which have both a geometry of domain time-varying surface. In this case two possible interpretations can be given to the topological predicate, depending on whether it must be satisfied either for at least one instant or for every instant belonging to both time extents of the varying geometries. Applied to the example of Fig. 1, this means that relationship Located only accepts instances that link a district and an avalanche zone such that their geometries intersect for at least one instant or for every instant belonging to both lifespans. When defining the relationship type, the designer has to specify which interpretation holds.

5 Multi-representation Modeling

Databases store representations of real-world phenomena that are of interest to a given set of applications. However, while the real world is supposed to be unique, its representation depends on the intended purpose. Thus, each application has a peculiar perception of the real world of interest. These perceptions may vary both in terms of what information is to be kept and in terms of how the information is to be represented. Fully coping with such diversity entails that any database element may have several descriptions, each one associated to the perceptions it belongs to. These multiple descriptions are called the *representations* of the element. Both metadata (descriptions of objects, relationships, attributes, is-a links) and data (instances and attribute values) may have multiple representations. There is a bidirectional mapping linking each perception to the representations perceived through this perception.

Classic databases usually store for each real-world entity or link a unique, generic representation, hosting whatever is needed to globally comply with all application perceptions. An exception exists for databases supporting generalization hierarchies, which allow storing several representations of the same entity in increasing levels of specificity. These classic databases have no knowledge of perceptions: Applications have to resort to the view mechanism to define data sets that correspond to their own perception of the database. Instead, MADS explicitly supports multiple perceptions for the same database. A *multi-perception database* is a database allowing users to store one or several representations for each database element, and records for each perception the representations it is made up.

Geographical applications have strong requirements in terms of multiple representations. For example, cartographic applications need to keep multiple geometries for each object, each geometry corresponding to a representation of the extent of the object at a given scale. Multiscale data/representations are needed as there is still no complete set of algorithms for cartographic generalization, i.e. the process to automatically derive a representation at some less detailed resolution from a representation at a more precise resolution.

In MADS, each perception has a user-defined identifier, called its *perception stamp*, or just *stamp*. In the sequel, perception stamps are denoted as s1, s2, ... sn. From data definitions (metadata) to data values, anything in a database (object type, relationship type, attribute, role, instance, value) belongs to one or several perceptions. Stamping an element of the schema defines for which perceptions the element is relevant. In the diagrams, e.g. Fig 3, the line identified by the ☞ icon defines the set of perceptions for which this type is valid. Similarly, the specification of the relevant stamps is attached to each attribute and method definition.

There are two complementary techniques to organize multiple representations. One solution is to build a single object type that contains several representations, the knowledge of "which representation belongs to which perception" being provided by the stamps of the properties of the type. Following this approach, in Fig. 3 the designer has defined a single object type RoadSegment, grouping two representations, one for perception s1 and one for perception s2. An object or relationship type is *multi-representation* if at least one of its characteristics has at least two different representations. The characteristic may be at the description level (e.g., an attribute with different definitions) or at the instance level (i.e., different sets of instances or an instance with two different values).

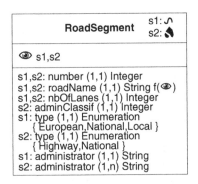

Fig. 3. An illustration of a bi-representation type, defined for perceptions s1 and s2

The alternative solution to organize multiple representations is to define two separate object types, each one bearing the corresponding stamp(s) (cf. Fig. 4). The knowledge that the two representations describe the same entities is then conveyed by linking the object types with a relationship type that holds a specific *inter-representation* semantics (indicated by the ⊕ icon). In the example Fig. 4, the same real-world road segment is materialized in the database as two object instances, one in RoadSegment1 and one in RoadSegment2. Instances of the relationship type Corresponds tell which object instances represent the same road segment.

The actual representation of instances of multi-representation object types changes from one perception to another. In the object type RoadSegment of Fig. 3 the spatial extent is represented either as a surface (more precise description, perception s2) or as a line (less precise description, perception s1) depending on resolution. Furthermore, perception s1 needs attributes number, roadName, numberOfLanes, type, and administrator. Perception s2 needs attributes number, roadName, numberOfLanes, adminClassification, type, and administrator. The type attribute takes its values from predefined sets of values, the sets being different for s1 and s2. Several administrators for a road segment may be recorded for s2, while s1 records only one. While the road segment number and the number of lanes are the same for s1 and s2, the name of the road is different, although a string in both cases. For instance, the same road may have name "RN85" in perception s1 and name "Route Napoléon" in s2. We call this a *perception-varying attribute* identified by the f(◉) notation. An attribute is perception-varying if its value in an instance may change from one perception to another. A perception-varying attribute is a function whose domain is the set of perceptions of the object (or relationship) type and whose range is the value domain defined for this attribute. These attributes are the counterpart of space-varying and time-varying attributes in the space and time modeling dimensions. Stamps may also be specified at the instance level. This allows defining different subsets of instances that are visible for different perceptions. For example, in the RoadSegment type in Fig. 3 it is possible to define instances that are only visible to s1, instances that are only visible to s2, and instances that are visible to both s1 and s2.

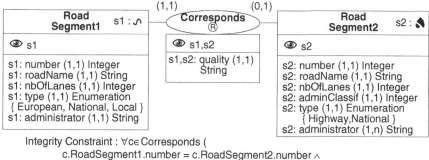

Fig. 4. The RoadSegment type (from Fig. 3) split into two mono-representation object types and an inter-representation relationship type

Relationship types are as dependent on perception as object types are. Therefore they can be multi-representation, like object types. Their structure (roles and association/multi-association kind) and semantics (e.g., topology, synchronization) may also have different definitions depending on the perception. For example in Fig. 1 it may be the case that the designer defines the relationship Occurs as 1) a topological and synchronization constraining relationship type for a perception s1, and 2) a plain relationship without any peculiar semantics or constraint for perception s2. A relationship type may have different roles for different perceptions. For example, in Fig. 1 Observes could be perceived in a representation s1 as a binary relationship between an observer and an avalanche event, while a perception s2 sees the same observation as a ternary relationship involving also the surveyor who has validated the observation.

Inter-representation relationship types, like any relationship type, belong to perceptions, and, although this is not likely to frequent happen, they also may have different representations according to the perceptions. In Fig. 4, the inter-representation Corresponds relationship type belongs to perceptions, s1 and s2, and has a unique representation. In particular, its attribute quality, which describes how well the two road segments, one from RoadSegment1 and one from RoadSegment2, correspond to each other[2], is shared by the two perceptions.

6 Conclusions and Future Work

Over the last decades conceptual modeling has been fundamental for improving our understanding about real-world phenomena as well as their implementation into computer systems. However, the globalization of the world as well as the rapidity at which social and economical changes occur calls for richer conceptual modeling approaches that can accumulate a larger variety of data as well as its evolution in time and in space, with detailed knowledge on change processes. Because of the many parties involved in complex data analyses, data representations must be flexible enough to allow recording many perspectives in view of exploring and managing alternative uses while keeping them interrelated so that more global analyses can also be performed. All of these features are advocated as important components of the active conceptual modeling paradigm that aims towards data management of the future.

In this paper we have shown how the MADS conceptual model can be used to capture the spatial and temporal characteristics of real-world phenomena, which are essential for their understanding in a dynamic geo-wide context. Further, we have shown how MADS allows keeping different perspectives of the same phenomenon. This is another essential requirement since complex phenomena must be analyzed under multiple perspectives, e.g., corresponding to economic, social, political, or environmental issues. Data analyses are in particular supported by data warehouses that aim at enabling strategic decision-making. Thus, a natural follow-on on active data modeling is active data warehousing, an emerging new research direction. A MADS companion and ongoing work is exploring a MADS inspired data model for data warehousing of spatial and temporal data [8].

MADS has already been used in many real-world applications, including the natural risk management application sketched in this paper. It thus provided fundamental support for decision making in these applications. Being conceptual it allows designers to focus on the issues at stake without being bothered by implementation constraints when designing and when manipulating the database. Further, specific CASE tools (available at http://cs.ulb.ac.be/mads_tools/) allow the

---

[2] The instances of the Corresponds relationship type are the result of a spatial matching process that looks for corresponding road segments in RoadSegment1 and RoadSegment2 by comparing their geometries. Roughly, the matching predicate means: Is the geometry of this road segment of RoadSegment1 topologically inside the geometry of that road segment of RoadSegment2?

translation of the data definition and manipulation languages into the languages provided by current implementation platforms.

The clean orthogonality we cared to follow in building the MADS approach is in our opinion what makes the real quality of the data model we propose. Orthogonality is the best way to provide maximum expressive power while keeping maximum simplicity in the constructs of the model. Needless to say, users' understanding of MADS and their ability to learn it and rapidly use it even in complex applications has been our greatest satisfaction. On these same premises, it will be easy to extend MADS to include more conceptual perspectives on additional modeling dimensions, e.g. uncertainty, multimedia, and movement. These are on our research agenda.

References

1. Chen, P.P., Thalheim, B., Wong, L.Y.: Future Directions of Conceptual Modeling. In: Chen, P.P., Akoka, J., Kangassalu, H., Thalheim, B. (eds.) Conceptual Modeling. LNCS, vol. 1565, pp. 287–301. Springer, Heidelberg (1999)
2. Parent, C., Spaccapietra, S., Zimányi, E.: Conceptual Modeling for Traditional and Spatio-Temporal Applications: The MADS Approach. Springer, Heidelberg (2006)
3. Parent, C., Spaccapietra, S., Zimányi, E.: The MurMur Project: Modeling and Querying Multi-Represented Spatio-Temporal Databases. Information Systems 31(8), 733–769 (2006)
4. Chen, P.P., Wong, L.Y.: A Proposed Preliminary Framework for Conceptual Modeling of Learning from Surprises. In: ICAI 2005. Proceedings of the International Conference on Artificial Intelligence, pp. 905–910. CSREA Press (2005)
5. Khatri, V., Ram, S., Snodgrass, R.: Augmenting a Conceptual Model with Geospatiotemporal Annotations. IEEE Transctions on Knowledge and Data Engineering 16, 1324–1338 (2004)
6. Pelekis, N., Theodooulidis, B., Kopanakis, I., Theodoridis, Y.: Literature review of spatio-temporal database models. The Knowledge Engineering Review 19, 235–274 (2004)
7. Bédard, Y., Bernier, E.: Supporting Multiple Representations with Spatial Databases Views Management and the Concept of VUEL. In: Proceedings of the Joint Workshop on Multi-Scale Representations of Spatial Data, ISPRS WG IV/3, ICA Comm. on Map Generalization (2002)
8. Malinowski, E., Zimányi, E.: Designing Conventional, Spatial, and Temporal Data Warehouses: Concepts and Methodological Framework. Springer, Heidelberg (to appear, 2007)

Postponing Schema Definition:
Low Instance-to-Entity Ratio (LItER) Modelling

John F. Roddick, Aaron Ceglar, Denise de Vries, and Somluck La-Ongsri

School of Informatics and Engineering
Flinders University,
P.O. Box 2100, Adelaide, South Australia 5001
roddick@infoeng.flinders.edu.au

Abstract. There are four classes of information system that are not well served by current modelling techniques. First, there are systems for which the number of instances for each entity is relatively low resulting in data definition taking a disproportionate amount of effort. Second, there are systems where the storage of data and the retrieval of information must take priority over the full definition of a schema describing that data. Third, there are those that undergo regular structural change and are thus subject to information loss as a result of changes to the schema's information capacity. Finally, there are those systems where the structure of the information is only partially known or for which there are multiple, perhaps contradictory, competing hypotheses as to the underlying structure.

This paper presents the *Low Instance-to-Entity Ratio* (LItER) Model, which attempts to circumvent some of the problems encountered by these types of application. The two-part LItER modelling process possesses an overarching architecture which provides hypothesis, knowledge base and ontology support together with a common conceptual schema. This allows data to be stored immediately and for a more refined conceptual schema to be developed later. It also facilitates later translation to EER, ORM and UML models and the use of (a form of) SQL. Moreover, an additional benefit of the model is that it provides a partial solution to a number of outstanding issues in current conceptual modelling systems.

1 Introduction

1.1 Systems Issues in Conceptual Modelling

The development of a conceptual schema commonly forms an important aspect of an information system's design. In this phase the structure of the system in terms of the relationships between objects, their attributes and constraints are established to the agreement of user and designer. Modelling techniques, such as ER/EER [1,2], NIAM/ORM [3,4] and UML [5], have been developed, extended and deployed effectively for this purpose over a number of years. Such systems fit well into the common software engineering frameworks that establish a firm design for the database before any data is stored [6].

P.P. Chen and L.Y. Wong (Eds.): ACM-L 2006, LNCS 4512, pp. 206–216, 2007.
© Springer-Verlag Berlin Heidelberg 2007

Some classes of information system, however, are not well served by these common techniques. These include:

- systems for which the number of entity-types is large in comparison with the number of instances stored. For these systems, the overhead of conceptual modelling can be high and can lead to short-cuts such as the aggregation of inappropriate entity-types.

- systems in which the immediate storage of data is the priority with the organisation of that data a secondary issue. For some systems, a mechanism to collect and store data is required more rapidly than the database design phase will permit. This includes systems designed rapidly in response to an immediate need [7,8].

- systems that undergo substantial structural change. While schema conversion, even those in which the schema's information capacity changes, does not always result in a loss of information[1], systems that regularly undergo change commonly lose information. Such systems include those used for hypothesis creation such as scientific databases [10], criminological systems [11] and *ad hoc* models established to track evolving phenomena.

- situations where the structure of the information is only partially known or where there are multiple, sometimes competing (although equally valid) models of the same data. While XML can handle semi-structured schema in which instances may possess varying structure, the overall schema is still largely formalised. However, we deal here with systems where the existence of different entity-types and the attributes they possess are largely unknown or where there is no agreement on the structure. Such systems include those that aim to handle empirical evidence in which the overall structure may be changed as ideas are developed and the evidence may still be in the process of being discovered. For these sorts of system, in any conflict between data and schema, it cannot be assumed that the schema is correct and that the data is therefore wrong. Sometimes it is the schema which is the component requiring change.

With effort, the traditional forms of conceptual modelling can handle these types of system although the overhead and side effects of doing so are often excessively high. In practice, the conceptual modelling step (and as a consequence the use of a DBMS) is often side-stepped because of the overhead involved resulting in systems missing the functionality able to be offered by databases.

In this paper we outline a new approach – LItER modelling – which lends itself to such systems. The common schema can be recorded in a database environment and, if required and with some caveats, can be translated to a conventional model once the confounding aspects listed above have dissipated.

---

[1] As discussed in other work [9] the limits for *practical* schema versioning in a database $\mathfrak{D}$ are that $S_1 \overset{p}{=} S_2$ iff $I'(\mathfrak{D}|S_1) \rightarrow I'(\mathfrak{D}|S_2)$ is bijective where $I'(\mathfrak{D}|S_n)$ is the set of all instances of S_n inferrable from $\mathfrak{D}$ given the constraints of S_n.

1.2   Data Issues in Conceptual Modelling

As well as the classes of system outlined in Section 1.1, there are other problems common in the modelling and implementation of even conventional systems. In order to provide some illustration of this, we provide here a motivating example based around the part-subpart and the supplier-part-project problems.

The ABC Company manufactures three types of widget - widayes, which are always blue regardless of who makes them, widbees, which the ABC Company paints blue, and widseas which are by default black but which can be painted according to the project on which they are used. The XYZ Company manufactures widayes, green widbees and has recently made a test batch of red widdees that are as yet unused. widayes are not only sold by themselves but are also used to make widseas. widayes are also known as ayewids.

Some of the problems illustrated by this example and that currently cause some concern include the following:

– It is often the case that collections of objects must be treated in the same manner as the objects themselves, often transitively, sometimes recursively. For example, if a batch of widayes are found to be defective then there may also be some widseas that also need to be recalled. This is particularly the case where groups are referred to in place of individuals (either through metonyms, holonyms or hypernyms).
– Attribute values are often provided to the system in ways that are not directly comparable despite conforming to the domain's type definition. For example, widayes may be described as *blue, dark blue, x3333cc, royal blue, PMS286* and so on. Synonyms, such as widayes and ayewids in the example, despite being relatively common, are not well accommodated. While data coercion is sometimes possible, this is not always a solution as the provenance and integrity of the original data may need to be maintained.
– In many conventional modelling techniques (such as EER), a relation formed from an *n*-ary or binary many-to-many relationship must, for reasons of

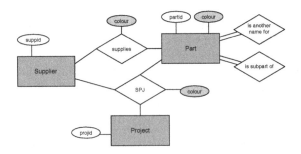

Fig. 1. Example Schema for Parts Example

entity integrity, have a stored instance for all its associated entities. In some cases this is either not possible or not desirable and the common practice is either to deform the model to suit a small fraction of the instances or, more commonly, to create *dummy* or default codes to circumvent this constraint. For example, in our example, since widayes only come in blue there may be little need to record the supplier or the project but for widbees and widseas the project must be recorded if the colour is needed. Moreover, what about the sample batch of red widdees? How is the colour to be stored? Classical modelling would required colour to be recorded in two (or more) places in the model, for example, in a schema such as that depicted in Figure 1.

1.3   Late Binding the Conceptual Modelling

The storage of data followed by conceptual model creation requires a different position to be adopted in that a generic or common conceptual model must exist for the initial data storage in the absence of the more specialised model. However, having established a common conceptual model, specialisations to that model can be developed incrementally through the testing and imposition of constraints.

For example, consider a scenario in which a University's student and faculty data is stored in a common storage structure (ignore for the moment the details of that storage structure). In the absence of any specialised schema, reference to entities and attributes must be phrased in terms of the structures provided by the common data model.

Over time, constraints could be tested and added providing more specialisation and eventually providing a level of structure consistent with a convention conceptual schema. For example, we might test, for example, through the discovery of induced dependencies [12], whether a doctoral student has exactly one supervisor and if so, and if such a constraint is considered sensible, it could be added. Labels could also be added so that conventional query languages such as SQL can function.

There are however, a few additional advantages to this approach:

– multiple, perhaps conflicting, structures can be held allowing, for example, transition between structures or the development of hierarchies of schema.
– multiple modelling paradigms can be used. For example, an EER model can be superimposed to provide a schema view while graph-oriented structures could be tested between data elements.
– having a common conceptual schema provides the ability to construct general utilities as well as facilitating schema integration.
– a common conceptual model lends itself to data mining as an *a priori* defined structure will not mask hidden associations.

The issue is to define a common schema which is flexible enough to hold all data but simple enough not to incur the overhead experienced when creating a full conceptual model. We argue that the proposed LItER modelling method

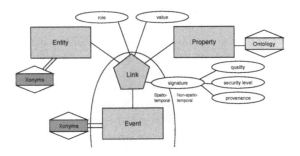

Fig. 2. LItER Schema

provides a common structural model capable of accommodating data derived from a variety of situations but which remains sufficiently structured to retain the semantics of the data.

2   LItER Modelling

The LItER approach has been developed as a consequence of a practical industry need to generate systems rapidly where the nature of the system is not known in advance but where some preparation (for example, the collection of meso-data and/or ontologies) can be undertaken. Such areas include national security, natural disaster and large-scale incidents where the details are unlikely to be known (at least not in detail) in advance. The LItER approach is to use a common schema discussed in Section 2.1 and shown in Figure 2, embedded within an overarching architecture discussed in Section 2.2 and shown in Figure 3.

2.1   The LItER Schema

The schema consists of three primary meta-entity-types, Entities, Properties and Events, together with a ternary Link (a form of polymorphic (overloaded) relationship-type). Events and Links are temporally referenced. Specifically, the model consists of the following components:

Entities: These represent objects in the model. This includes not only elementary objects such as those that might be represented by strong and weak entities in an EER Model but also components and aggregations of entities. If data is obtained from multiple sources, the same entity may also be recorded more than once (linked by a synonym link). The only attribute directly recorded is the entity's identifier (which might be user or system supplied); all other attributes being recorded through reference to a property.

Properties: These allow the description of properties that can be associated with either an entity or an event. A property may be associated with an ontology which can be used by the constraint manager/hypothesis checker or by the query language as appropriate.

Events: These allow the recording of spatio-temporally referenced happenings. Once again, the only attribute directly recorded is the event's identifier.

Links: The link allows entities, properties and events to be combined. In LItER, we use an overloading form of polymorphism to specify that a link can occur between all or any of Entities, Events or Properties. Moreover, more than one of each can participate in a link with the sole requirement being that a link must include at least two identifying instances. Thus a link can allow:

- one or more entities to be associated with a describing property, optionally with a *value* of that property. Note that the value may also be provided as a simple formula which may include a variable (such as > 25 or $P(Age).E(Christopher) + 2$). This allows facts such as ... *is over 25* or ... *is 2 years older than Christopher* to be recorded[2].
- one or more events to be associated with a describing property, optionally with a *value* for that property.
- a link between an entity (or entities) and an event such that an entity's *role* in the event can be recorded.
- a relationship between two entities.

Links have a number of optional predefined attributes as follows:

Value - providing a qualification for the relationship being described. For example, Entity Mary linked to Property Nationality might have the value Australian.

Role - providing a qualification for the relationship being described. For example, Entity Luke linked to Event Phone_Call might have the role Caller.

Quality - provides a measure of confidence (such as a probability) to the link;

Security level - provides a mechanism for restricting access to data;

Provenance - provides a mechanism to record the owner or source of the data.

Ontologies and Xonyms: These are an integral part of the model.

Ontologies: Full ontologies (which for this purpose we define as complex domain structures) are associated with properties. Such ontologies are similar to the mesodata concept discussed by de Vries [13,14].

Xonyms: There are a variety of common binary references that are used and understood widely. Xonyms allow such common linkages between objects to be recorded more simply. For example, a Person can be found to *be the same as* another. A Meeting *is a form of* Communication and so on. In these cases *synonym* and *hypernym* references would be created resp.

---

2 Specifically, we allow first order formulae over constants or variables participating in the link plus those accessible through graph traversal. For example, $P(Age).E(Christopher) + 2$ references the value associated with link between the entity with entity identifier *Christopher* and the Property *Age* while $max(P.(Age).E(*).P(InDept[Consultant].Sales))$ returns the maximum *Age* of all entities with a link to Property *InDept* with a value of *Sales* and a role of *Consultant*.

Other Xonyms include acronyms, holonyms (and meronyms), hypernyms (and hyponyms), metonyms, pseudonyms, and synonyms. The alternative, merging Entities, would result in a loss of both information and provenance, particularly if the reason for the merge was later discredited.

While in many cases it is not difficult to extend most query languages and data mining routines to be able to understand the semantics of ontologies and Xonyms, by making them part of the model it is possible that some systems need not have to do so. For example, association mining routines might be given their input data with all synonyms resolved.

2.2   The LItER Architecture

While the LItER model is independently useful, an overarching architecture has been developed which maximises the benefits of the model (shown in Figure 3). Some of the important points are discussed below.

Analysis Routines. Four sets of routine are made available to the user:

   Predicate Definition and Data Dictionary. These provide a resource to allow easy reference to data items.

   Query Languages. It is possible to provide a form of SQL which resolves the terms provided by reference to the Predicate Definitions and Data Dictionary. For example, the query:

```
SELECT  EmpNAME
FROM    EMPLOYEE
WHERE   EmplAGE < 25;
```

   might (depending on how the data was organised) be resolved by the following definitions:

```
EMPLOYEE  ::= {E(*).P(WorkFor)}
EmpNAME   ::= P.(HasName).EMPLOYEE
P.HasAge  ::= V.(WasBorn).E(*) - TODAY()
EmpAGE    ::= P.(HasAge).EMPLOYEE
```

   The first statement creates a set of instances of Entity. The second returns the value for the link to the Property *HasName*. The third creates a virtual property of *HasAge* which exists for all instances of Entity with a link to an Event of type *WasBorn*. The last returns that value for all instances of *EMPLOYEE*.

   Constraint Manager / Hypothesis Checker. This allows the creation of structures that are either used to constrain the data or as putative hypotheses that can be checked against the data.

   Data Mining Routines. One of the drawbacks of many data mining systems is the lack of reuseability caused, in part, by changes in the manner in which data are stored. LItER accommodates data mining routines which now have access to a common schema. In particular, graph mining [15] and association mining [16] have been found to be useful. Importantly, these routines are generally generic and independent of the data.

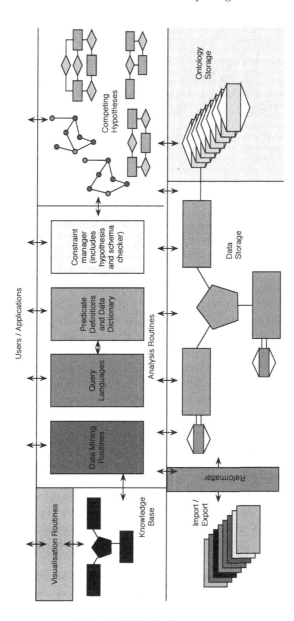

Fig. 3. LItER Architecture

For example, a graph mining routine that seeks to characterise modes of communication between actors can also be applied, given the same LItER schema, to any other graph in which the entities are linked through some event, for example, the transmission of infection.

A knowledge base. In most cases the knowledge base uses the same LItER architecture. Importantly, visualisation routines can operate over the knowledge base and this can take advantage of multiple runs and different forms of data mining.

A general ontology storage area. This can be populated independently of the data storage (ie. it can be made ready in advance of any intended use). To date, the ontologies used have been restricted to complex structures of attribute values the mesodata concept [13][3].

3   Discussion

The model has a number of important modelling characteristics.

1. Because the objects in a system can play many different roles, entities are not directly typed. Instead their types are recorded by virtue of the properties (and perhaps the property-value pairs) they hold. Thus property owners may be identified by being linked to the property *is an owner* (cf. the category concept of Elmasri [17]), Australians by having the property-value pair of *having nationality* with value *Australian*. Significantly, this allows the creation of heterogeneous sets - being *Australian* is, of course, a property that could be assigned to more than just people.

2. As the data stored in such a system is often used for hypothesis creation, the model must also allow for temporal auditing and probabilistic reasoning. Moreover, such systems often obtain data from various sources and therefore not only must the provenance of the data be recorded but also, as far as practicable, the format and content of the data must be maintained (and thus data matching is an important component of the system).

3. Relationships, and their cardinalities, are induced rather than explicitly stored. In many cases, the choice to make a property of an object an attribute rather than a relationship to an entity-type representing the concept is largely dependent on the data available (qv. the semantic ambiguity as discussed by Wand, Storey and Weber [18]). Consider the example in Figure 4 in which either an attribute or a relationship-type and entity-type are used. In this respect, the LItER model mirrors the bottom-up approach used by ORM.

4. Unlike the EER Model, relationships (renamed here as links to avoid confusion) are polymorphic; a variable number of Entities can be required to provide the key for a link. For example, the property *has colour* may be specified with a *part number* and the *project* and/or with just a *part number*. In LItER, this polymorphic use of relationships is allowed subject to there not being a constraint forbidding it.

---

[3] Mesodata aims to add semantic capability by providing greater semantics to the domain of an attribute by allowing attributes to be defined over complex domain structures. For example, while the code for a disease might be defined as CHAR(5), disease codes exist within an agreed international classification (such as ICD10), a tree-structure that relates diseases and other observations by group.

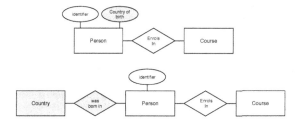

Fig. 4. Two design choices

5. The recording of constraints used to gradually refine the model can be used to either validate data (either before or after DBMS commit) or to validate the schema, and to determine contradictory information. Moreover, constraints can take the form of hypotheses, in which ideas can be tested and the extent of missing information ascertained.

Note that the systems for which the LItER model is most suited do not necessary have low volumes of data. What distinguishes the model for these systems is that the information held is diverse with some coming from large databases but with the structure of other information being more or less specific to one or a few entities only.

4 Conclusions and Further Work

The LItER model and architecture is being developed as a result of genuine industry requirements and is being applied in both a defence and health environment. Interestingly, as has been noticed in some data mining research [19], a cascade effect has been noticed in which the more data is added, the greater the overall connectivity between data objects is experienced.

There has been considerable discussion as to the role of Reiter's closed world assumption [20] and whether it can be assumed to hold in the context of some systems. This, and the accommodation of negative information is currently the subject of further investigation.

While the LItER model does not fulfil all conceptual modelling needs, (we restrict ontologies to conform to the concepts of mesodata domains for example), its development is generating substantial interest. In particular, its ability to rapidly provide areas for subsequent investigation and its ability to quickly bring together data from a variety of data sources has been of great interest.

References

1. Chen, P.P.S.: The entity-relationship model - toward a unified view of data. ACM Transactions on Database Systems 1, 9–36 (1976)
2. Thalheim, B.: Entity-Relationship Modeling: Foundations of Database Technology. Springer, Berlin (2000)

3. Halpin, T.: Object-role modeling (ORM/NIAM). In: Bernus, P., Mertins, K., Schmidt, G. (eds.) Handbook on Architectures of Information Systems, pp. 81–101. Springer, Berlin (1998)

4. Verheijen, G., van Bekkum, J.: NIAM: an information analysis method. In: IFIP WG8.I Working conf. Information Systems Design Methodologies: a comparative review, North Holland Publishing, Netherlands (1982)

5. Booch, G., Jacobson, I., Rumbaugh, J.: Unified modelling language user guide, 2nd edn. Addison Wesley Professional, Reading (2005)

6. Sommerville, I.: Software Engineering, 8th edn. Addison-Wesley, Boston, MA, USA (2006)

7. Chen, P.P.S.: Suggested research directions for a new frontier - active conceptual modeling. In: Embley, D.W., Olivé, A., Ram, S. (eds.) ER 2006. LNCS, vol. 4215, pp. 1–4. Springer, Heidelberg (2006)

8. Roddick, J.F., Ceglar, A., de Vries, D.: Towards active conceptual modelling for sudden events. In: Grundy, J., Hartmann, S., Laender, A., Maciaszek, L., Roddick, J. (eds.) 26th International Conference on Conceptual Modeling (ER 2007) (Posters). CRPIT, Auckland, New Zealand, vol. 83, pp. 203–208. ACS (2007)

9. Roddick, J.F., de Vries, D.: Reduce, reuse, recycle: Practical approaches to schema integration, evolution and versioning. invited keynote address. In: Grandi, F. (ed.) ECDM 2006. LNCS, vol. 4231, pp. 209–216. Springer, Heidelberg (2006)

10. Shoshani, A., Wong, H.K.T.: Statistical and scientific database issues. IEEE Transactions on Software Engineering 11, 1040–1047 (1985)

11. Chen, H., Zeng, D., Atabakhsh, H., Wyzga, W., Schroeder, J.: Coplink: managing law enforcement data and knowledge. Communications of the ACM 46, 28–34 (2003)

12. Roddick, J.F., Craske, N.G., Richards, T.J.: Handling discovered structure in database systems. IEEE Transactions on Knowledge and Data Engineering 8, 227–240 (1996)

13. de Vries, D., Rice, S., Roddick, J.F.: In support of mesodata in database management systems. In: Galindo, F., Takizawa, M., Traunmüller, R. (eds.) DEXA 2004. LNCS, vol. 3180, pp. 663–674. Springer, Heidelberg (2004)

14. de Vries, D.: Mesodata: Engineering Domains for Attribute Evolution and Data Integration. PhD thesis, Flinders University (2006)

15. Chakrabarti, D., Faloutsos, C.: Graph mining: Laws, generators, and algorithms. ACM Computing Surveys 38 (2006)

16. Ceglar, A., Roddick, J.F.: Association mining. ACM Computing Surveys 38 (2006)

17. Elmasri, R., Weeldreyer, J.A., Hevner, A.R.: The category concept: an extension to the entity-relationship model. Data and Knowledge Engineering 1, 75–116 (1985)

18. Wand, Y., Storey, V.C., Weber, R.: An ontological analysis of the relationship construct in conceptual modeling. ACM Transactions on Database Systems 24, 494–518 (1999)

19. Spencer, J.: The Strange Logic of Random Graphs. Springer, Heidelberg (2001)

20. Reiter, R.: On closed world databases. In: Gallaire, H., Minker, J. (eds.) Logic and Databases, pp. 55–76. Plenum Press, New York (1978) reprinted In: Mylopoulos, J., Brodie, M.L. (eds.) Artificial Intelligence and Databases, pp. 248–258. Morgan Kaufmann

Research Issues in Active Conceptual Modeling of Learning: Summary of Panel Discussions in Two Workshops (May 2006) and (November 2006)

Peter P. Chen[1,*], Leah Wong[2,**], Lois Delcambre[3,***], Jacky Akoka[4], Arne Sølvberg[5], and Raymond Liuzzi[6]

[1] Computer Science Department, Louisiana State University
Baton Rough, LA 70803, U.S.A.
[2] Space and Naval Warfare Systems Center San Diego
Code 246214, 53560 Hull Street
San Diego, CA 92152, U.S.A.
[3] Computer Science Department, Maseeh College of Engineering and Computer Science
Portland State University
Portland, OR 97207, U.S.A.
[4] Conservatoire National des Arts et Métiers and INT, France
[5] Dept. of Computer and Information Science (IDI)
Norwegian University of Science and Technology (NTNU)
N-7491 Trondheim-NTNU, Norway
[6] (Formerly U.S. Air Force Research Lab.), Raymond Technologies
14 Bermuda Rd., Whitesboro, NY 13492, U.S.A.
pchen@lsu.edu, leah.wong@navy.mil, imd@cs.pdx.edu,
asolvber@idi.ntnu.no, akoka@cnam.fr, raymondtechnologies@msn.com

Abstract. The SPAWAR Systems Center (SSC San Diego) of the U.S. Navy hosted two workshops on Active Conceptual Modeling of Learning (ACM-L). The first workshop was held at SSC San Diego on May 10-12, 2006 to introduce the Science &Technology (S&T) Initiative and identify a Research and Development agenda for the technology development investigation. Eleven invited researchers in Conceptual Modeling presented position papers on the proposed S&T Initiative. The second workshop was held on November 8, 2006 at the 25th International Conference on Conceptual Modeling, ER 2006, 6-9 November 2006, in Tucson, Arizona. Complementary to the May Workshop, the November workshop was a forum for the international researchers and practitioners to present their papers as a result of a call for papers and to exchange ideas from various perspectives of the subject. This paper describes research issues identified by participants from the two ACM-L workshops.

* The research of this author was partially supported by National Science Foundation grant: NSF-IIS-0326387 and Air Force Office of Scientific Research (AFOSR) grant: FA9550-05-1-0454.
** Contributions to this article by co-author Leah Wong constitute an original work prepared by an officer or employee of the United States Government as part of that person's official duties. No copyright applies to these contributions.
*** The research of this author was partially supported by National Science Founation grant: NSF-IIS-0534762.

P.P. Chen and L.Y. Wong (Eds.): ACM-L 2006, LNCS 4512, pp. 217–225, 2007.
© Springer-Verlag Berlin Heidelberg 2007

1 Introduction

Information technology, which is viewed as the cornerstone for our nation's technology infrastructures, has made a tremendous impact on all aspects of our society. Central to this cornerstone is the process of conceptual modeling. In looking back to the past three decades since the inception of the Entity-Relationship (ER) Model originated by Peter Chen in 1976 [1], the ER modeling approach has gained worldwide acceptance in database design, software engineering, information system development, and system modeling/specifications. In the past three decades, many papers of extending and modifying the original ER model have been presented in the annual International Conference on Conceptual Modeling [2].

The conventional/traditional conceptual modeling concentrates on modeling the static views (i.e., a snapshot) resulting in fixed representation of the real world domain. The original goal of a database was to model some aspects of the real world of interest. The purpose of traditional data modeling is to help us better understand a specific real-world domain and enhance communication among ourselves. With the advent of the Internet, conceptual modeling is shifting to the proposed active paradigm. Information management, which has a new meaning in the context of the Internet, calls for new modeling techniques.

We are entering into the information age where global awareness is a necessity. We observed an increased demand for supporting complex applications that require integration of historical, active information, and past experiences in order to accomplish global and long-term situation awareness and monitoring. Recent incidents (e.g., September-11, tsunami, Katrina, etc.) forced us to look back to the past changes from a holistic perspective in order to analyze crises, abnormal behaviors in surprises, and unconventional events.

Increasing changes in the real world demand a shift in conceptualization and a new way of viewing reality and using the evolving knowledge. There are growing needs of traceability for the evolving and changing world state. There are also increasing needs for understanding relationships among changes, which may have significance to current world state. We envision Active Conceptual Modeling will be the next major development of conceptual modeling, which will allow for continual learning from past experience, including surprises, and potentially be useful for predicting future actions.

Complementary to other areas, which attempt to develop systems to accumulate knowledge based on past experience and analytical observations, this effort focused on relationships between past knowledge/data and current knowledge/data from different perspectives. We proposed a framework for active conceptual modeling.

Active conceptual modeling [3, 4] is a continual process that describes all aspects of a domain, its activities, and changes under different perspectives based on multi-perspective knowledge and human cognition. For any given time, the model is viewed as a multi-level / multi-perspective, high-level abstraction of reality.

Conventional conceptual modeling for database design is a simple case of active modeling. The active conceptual model will provide the necessary control and traceability for the domain by linking snapshots to form a dynamic and moving picture of the evolving world. This single dynamic model will integrate temporal and spatial entities, time-varying relationships, temporal events, dynamic classification of entities, and uncertainty. The model will help us understand relationships among state changes and to continually learn and make inferences by providing traceable lessons learned from past experiences, which could be used to predict future actions. The active model would therefore subsume the current databases and knowledge bases.

The technical challenges of active modeling of information are significant. To begin framing the problem, two workshops were organized by a group of active researchers in the field. The results pointed to several key threads of the research. The findings from the two workshops are described in the following sections. We conclude with a set of research issues as future directions.

2 The First ACM-L Workshop

The first workshop on ACM-L was held at SSC San Diego on May 10-12, 2006. It was attended by invited researchers/participants and SSC San Diego personnel.

2.1 Goal

The workshop was convened to accomplish the following goals:

- Introduce the Science and Technology (S&T) challenge
- Provide a forum for the exchange of ideas and research results of the proposed research and the impact in the Department of Defense (DoD) and commercial applications
- Identify a research and development (R&D) agenda for a technology development investigation.

The ACM-L effort aims to develop ACM-L technology for next generation learning-base system development. The effort focuses on enhancing our fundamental understanding of how to model continual learning from past experiences, and how to capture knowledge from transitions between system states. This understanding will enable us to provide traceable lessons learned to improve current situations, adapt to new situations, and potentially predict future actions. The goal is to provide a theoretical framework for the ACM-L. This framework is based on the ER approach and the human memory paradigm for developing a learning base to support complex applications such as homeland security; global situation monitoring; command, control, communication, computers, intelligence, surveillance, and reconnaissance(C4ISR); and cognitive capability development.

2.2 Summary of Findings

The technical sessions focused on the following areas and issues:

1. **Time and Events Issues:** Semantics of connections between time and events; characterization and classification of events; complexity of events with issues of composition, decomposition, splitting, merging and overlapping, etc.; temporal logic and granularity; the meaning of time between state and event, initiation,

detection, processing, timeframe for decision-making, and time in the future; and integrating time in conceptual modeling
Short Term: Temporal concepts and terminology for active conceptual modeling
Medium Term: Time-related event processing system
Long Term: Integrating temporal concept in active conceptual model

2. **Continuous Changes Issues:** Representing snapshots and time-varying attribute values as a means for modeling changes, computing delta between snapshots with different types of data, preserving time-varying data reflecting their multi-level and multi-perspective environment, ecological processes and operations, and methods for detecting changes in multimedia data
   Short Term: Recognizing changes in structured and semi-structured data
   Medium Term: Methods for detecting changes in multimedia data

3. **Understanding and Processing of "Stories" Issues:** Vagueness, uncertainty, inconsistency. What makes a set of events a "story"? Research in AI has dealt with unstructured data, but only with a few cases involving a few parameters. The approaches were informal and the interpretations often were not unique. Data have many forms, and establishing commonality is difficult. "Markers" have been used to extract and summarize information, and semantics of "markers" need to be defined more clearly. Conceptually processing "stories" is feasible, but the scalability issues of generating and processing" stories" may be difficult. Other issues are the verification of "stories," provenance issues that include semantics of provenance and automation, and acquisition of provenance.
   Short Term: Conceptually modeling "stories" in ER framework
   Medium Term: Generating and processing "stories"
   Long Term: Assessment and evaluation of impact in decision-making

4. **Events and Triggers Issues:** Representing workflow on top of the events, creating a trigger view on top of the conceptual view, rich triggers with spatial and temporal processes, and integrating flow processes with policies and regulations
   Short Term: Diagrammatic techniques for representing workflow and conceptual views
   Medium Term: Diagrammatic representation of workflow over events, triggers with spatial and temporal features
   Long Term: Flow processes with policies and regulations with entities

5. **Learning and Monitoring Issues:** Reference on learning includes knowledge discovery, inference, machine learning, case-based and analogical reasoning, and causality; generalization from instances to schema and schema evolution based on past experience or new data collected, the detection of "anomalous events;" whether new data are consistent with known "signature" concepts or rules; limitations and constraints need to be considered when an action taken depends on the time window and number of outliers; and the use of "sliding windows" for continuous processing.
   Short Term: Realizing rules and knowledge learning from instances
   Medium Term: Detecting some types of anomalous behavior
   Long Term: Continuous learning and knowledge discovery

6. **Schema Evolution and Version Management Issues:** Preserving historical changes with schema evolution, processing data under a past environment with rules and policies applicable at the time in the past, tracing the sources of the changes in the past and forecasting future trends, representing incremental and discrete changes with versions, and version migration and management.
   Short Term: Linking ER schema changes to XML
   Medium Term: Developing tools for data migration
   Long Term: Providing schema evolution in ER conceptual model

7. **Executable Active Conceptual Model and Database Management System Issues:** Creating executable conceptual model; supporting changes in conceptual model; generating and maintaining incremental executable versions; ensuring information and data semantic integrity; managing persistent storage for active data; providing transparency and traceability between users, applications, conceptual schema, and stored data.
   Short Term: Concepts and basic building blocks for executable conceptual model
   Medium Term: General architecture for ER database management system
   Long Term: ER database management system for storing and accessing active data.

3   The 2nd Workshop: The First International ACM-L Workshop

The second workshop (officially called the First International Workshop on ACM-L) was held on November 8, 2006 at the 25th International Conference on Conceptual Modeling, ER 2006, 6-9 November 2006, in Tucson, Arizona. It was a follow-up to the first ACM-L workshop held earlier in May 2006. As a result of the call for papers, the Program Committee received 26 submissions from 10 countries, and after rigorous refereeing, 11 papers were eventually chosen for presentation at the workshop. The workshop program was enriched by a keynote address given by Peter Chen, who introduced Active Conceptual Modeling as a new frontier of research. Over 50 people attended the workshop consisting of eleven paper presentations, a panel discussion, and open discussion.

3.1   Goal

The goals of this workshop were to introduce a new frontier of research: active conceptual modeling for learning, suggest research directions, and stimulate future research. The workshop brought together a wide range of international researchers sharing the common goal that active modeling based on the ER approach could provide an exciting new direction in the field of conceptual modeling.

The ACM-L objectives were to:

- Link snapshots to form a dynamic, moving picture of the evolving world with traceability
- Capture relationships among state changes to provide trends and lessons learned from past experience

- Represent and track changes to the underlying model for knowledge management and application support
- Provide a unified dynamic model integrating semantic, temporal, spatial, and uncertain relationships in formation and events
- Serve as a basis for executable user interface and active system implementation.

3.2 Summary of Findings

The paper presentations are described in this book volume. Peter Chen led a panel discussion on ACM-L following the paper presentations. The panel was composed of several distinguished researchers: Arne Sølvberg, Raymond A. Liuzzi, Stefano Spaccapietra, and Reind P. van de Riet. Peter Chen opened the panel by suggesting that modeling data and processes over time and space could generate learning. Each panelist discussed a variety of topics relating to ACM-L that included the effect of context awareness on ACM-L models, security concerns, and relationship to cognitive research for ACM-L. The key points of the opening remarks of these panelists are:

- Arne Sølvberg (Norwegian Institute of Technology, Norway) discussed modeling differences related to ACM-L: using work-flow for information system, problem solving in processes, modeling language, which includes time and events. We need to examine the essence of each of these approaches.
- Raymond A Liuzzi (Raymond Technologies) discussed ways in which DARPA's Cognitive Processing Research Program could benefit from ACM-L research. Specifically, he discussed how the Perceptive Assistant that Learns (PAL) Program could use ACM-L research to further learning in military applications.
- Stefano Spaccapietra (Ecode Polytechnique Federale Lausanne, Switzerland) emphasized the problem of context awareness, which is one of the most important investigations in Europe. He discussed the MADS conceptual model for capturing the spatial and temporal characteristics of real-world phenomena with multi-representation support and data context awareness related to ACM-L.
- Reind P. van de Riet (Professor Emeritus, Free University, Amsterdam, Netherlands) discussed security concerns of web service with respect to context-awareness for access control and learning from disasters related to ACM-L.

The ACM-L Workshop concluded with an hour discussion that centered on a number of issues triggered by Peter Chen's and the panelists' opening remarks. These opening remarks invoked lots of discussion on the relationships between ACM-L, active learning, and context learning, and the roles that ACM-L play in information systems and support for applications such as:

- How can we manage changes in learning?
- How can ACM-L react to surprises? Is there a way to change form one model to another one?
- What should be the core constructs for ACM-L? We should also consider modeling relationship of relationships and use event models to represent situations.

- How might we beneficially merge the role of AI and database theories within ACM-L? This generated discussion on how the Cyc system could be used as an example of model of learning based on its theories and sub-theories. This approach to learning might have application to ACM-L.

4 Future Directions

Among a broad spectrum of issues discussed, we have identified key technical areas and stressed the importance of developing an S&T team to address them. The following categories are based on results from the two workshops:

1. Integrating time, space, and perspective dimensions in a theoretical framework of conceptual models
   - ER theory
   - Core constructs in active conceptual model of learning
   - Mathematical framework for active conceptual models
   - Mapping of constructs among conceptual models
   - Multi-level conceptual modeling
   - Multi-perspective conceptual modeling
   - Aspects of conceptual modeling
   - Multi-media information modeling

2. Management of continuous changes and learning
   - Continuous knowledge acquisition
   - Information extraction, discovery, and summarization
   - Experience modeling and management
   - Learning from experience
   - Management of changes in knowledge representation
   - Representation of changes in knowledge
   - Transfer learning in time dimension
   - Lessons learned capturing

3. Behaviors of evolving systems – including model evolution, patterns, interpretation, uncertainty, and integration
   - Time and events in evolving systems
   - Situation monitoring (system-and user-level monitoring)
   - Reactive, proactive, adaptive, deductive capability in support of active behavior
   - Schema evolution and version management
   - Context awareness and modeling of context changes
   - Information integration and interpretation
   - Pattern recognition over a time period
   - Uncertainty management with respect to integrity
   - Combined episodic and semantic memory paradigm for structuring of historical information

4. Executable conceptual models for implementation of active systems
   - Dynamic reserve modeling
   - Storage management
   - Security
   - Architectures for information system based on the active conceptual model
   - Languages for information manipulation
   - User interface
   - Bench marking for Test & Evaluation.

5 Conclusion

Here, we use the main message of Peter Chen's keynote address at ER 2006 Conference [5] as the concluding remarks. In his keynote address, he presented the history of conceptual modeling in the past three decades from his original development of the ER model in1976 [1] to the follow-on joint efforts of researchers and practitioners. He classified the past three decades into the following phases:

(1) The first five years: the road toward acceptance
(2) The last twenty five years: related developments
(3) The present status: worldwide acceptance.

He stated that the ER modeling was triggered by the critical needs for unifying data views from top-down and bottom-up perspectives and integration of more data semantics. Entity and relationship are the most fundamental concepts used for Data/Knowledge Representation, database design, software engineering, information system development, data mining, and system modeling/specifications. In the past three decades, the ER modeling methodology has been the most widely used methodology in business database applications and government agencies. Most CASE tools support ER modeling including the Oracle Designer/2000, Computer Associates ERWIN, Sybase Power Designer, Microsoft Access, Visio, etc. The recent UML (Unified Modeling Language) reinforces the ER concepts. Object-Oriented (OO) modeling incorporates many ER concepts although current OO methodologies still need more general concepts of relationship. Data mining is an implicit way of constructing ER models from data to discover hidden relationships and/or the embedded ER models.

We envision Active Conceptual Modeling will be the next major development of conceptual modeling. It can help us understand the relationships of past events and make better decisions for future events. A refined ER model could be a feasible and viable solution. Active Conceptual Modeling can only be realized through technology integration (e.g. AI, software engineering, information/knowledge management, cognitive science, neuroscience, philosophy, history, etc.) An executable conceptual model is becoming a reality. We accomplished our goals by introducing the challenges of the new frontier, providing a forum for brainstorming issues of the initiative, reporting findings, and motivating further investigations.

References

1. Chen, P.P.: The Entity-Relationship Model: Toward a Unified View of Data. ACM Transaction on Database Systems 1(1), 9–36 (1976)
2. ER Conference main website: http://www.conceptualmodeling.org
3. Chen, P.P., Thalheim, B., Wong, L.Y.: Future Directions of Conceptual Modeling. In: Chen, P.P., Akoka, J., Kangassalu, H., Thalheim, B. (eds.) Conceptual Modeling. LNCS, vol. 1565, pp. 287–301. Springer, Heidelberg (1999)
4. Chen, P.P., Wong, L.Y.: A Proposed Preliminary Framework for Conceptual Modeling of Learning from Surprises. In: Proceedings of the International Conference on Artificial Intelligence (ICAI), pp. 905–910. CSREA Press (2005)
5. Chen, P.P.: Suggested Research Directions for a New Frontier–Active Conceptual Modeling. In: Embley, D.W., Olivé, A., Ram, S. (eds.) ER 2006. LNCS, vol. 4215, pp. 1–4. Springer, Heidelberg (2006)

Author Index

Lecture Notes in Computer Science

Sublibrary 3: Information Systems and Application, incl. Internet/Web and HCI

For information about Vols. 1– 4519
please contact your bookseller or Springer